ALGEBRA
A Computational
Introduction

Studies in Advanced Mathematics

JOHN SCHERK

Department of Mathematics
University of Toronto
Ontaria, Canada

ALGEBRA
A Computational
Introduction

CHAPMAN & HALL/CRC

Boca Raton   London   New York   Washington, D.C.

ON THE COVER: The pencil of conics associated with a quartic, drawn by *Mathematica* code written by Petra Menz.

Library of Congress Cataloging-in-Publication Data

Scherk, John.
    Algebra : a computational introduction / John Scherk.
        p. cm.—(Studies in advanced mathematics)
    Includes bibliographical references and index.
    ISBN 1-58488-064-3
    1. Algebra. I. Title. II. Series.
QA154.2 .S28 2000
512—dc21   00-031432
                                                                                      CIP

Visit the CRC Press Web site at www.crcpress.com

© 2000 by Chapman & Hall/CRC

No claim to original U.S. Government works
International Standard Book Number 1-58488-064-3
Library of Congress Card Number 00-031432
Printed in the United States of America   3 4 5 6 7 8 9 0
Printed on acid-free paper

Contents

Preface

This text is an introduction to algebra for undergraduates who are interested in careers that require a strong background in mathematics. It will benefit students studying computer science and physical sciences, who plan to teach mathematics in schools, or to work in industry or finance. The book is written for readers with a solid background in linear algebra. For the first 12 chapters, elementary operations, elementary matrices, linear independence, and rank are important. In the second half of the book, abstract vector spaces are used. Students will need to have experience proving results. Some acquaintance with Euclidean geometry is also desirable. In fact, a course in Euclidean geometry fits together well with the algebra in the first 12 chapters. But one can avoid the geometry in the book by simply omitting Chapter 7 and the geometric parts of Chapters 9 and 18.

The material in the book is organized linearly. There are few excursions from the main path. The only significant parts that can be omitted are those just mentioned, the section in Chapter 12 on $PSL(2, \mathbb{F}_p)$, Chapter 13 on abelian groups, and the section in Chapter 14 on Berlekamp's algorithm.

The first chapter is meant as an introduction. It discusses congruences and the integers modulo n. Chapters 3 and 4 introduce permutation groups and linear groups, preparing for the definition of abstract groups in Chapter 5. Chapters 8 and 9 are devoted to group actions. Lagrange's Theorem is detailed in Chapter 10 as an application. The Sylow theorems in Chapter 11 are proved following Wielandt via group actions as well. In Chapter 13, row and column reduction of integer matrices is used to prove the classification theorem for finitely generated abelian groups. Chapter 14 collects all the results about polynomial rings in one variable over a field that is needed for the Galois theory. I have followed the standard Artin–van der Waerden approach to the Galois theory. But I have tried to show where it comes from by introducing the Galois group of a polynomial as its symmetry group, that is, the group of permutations of its roots which preserves algebraic relations among them. Chapters 18, 19, 20, and 21 are applications of the Galois theory. In Chapter 20, I have chosen to prove only that the general equation of degree 5 or greater cannot be solved by taking roots. I have found that the correspondence between radical extensions and solvable Galois groups is often too sophisticated for undergraduates.

This book also shows students how software can be used intelligently in algebra. I feel that this is particularly important for the intended audience. There is a delicate philosophical point. Does a software calculation prove anything? This is not a simple question, and there does not seem to be a consensus among mathematicians regarding its answer. There are a few places in the text where a calculation does rely on software, for example, in calculating the Sylow 2-subgroups of S_8. The *Mathematica* notebooks corresponding to the software sections are available at the publisher's web site, **www.crcpress.com**, as are the equivalent *Maple* worksheets.

Some of the exercises are referred to later in the text. These have been marked with a bullet (•). There are exercises where the software is useful but not essential, and some where it is essential. However, I have deliberately not indicated which ones these are. Learning to decide when software is useful and when not, seems to me to be an important part of learning to use it.

I am grateful to many people for help with this book at various stages, in particular to Edward Bierstone, Imtiaz Manji, David Milne, Kumar Murty, Joe Repka, and Paul Selick. In discussions over the years, Ragnar Buchweitz has made many suggestions about teaching undergraduate algebra, for which I am most thankful. The section on quartics and the associated pencil of conics is one of several topics in the book suggested by him. The software was originally written with the help of George Beck, Keldon Drudge, and Petra Menz. The present version is due to David Milne. The software that produced the pictures of the Platonic solids in Chapter 7 was also written by George Beck.

J. S.

Chapter 1

Congruences

This is an introductory chapter. The main topic is the arithmetic of congruences, sometimes called "clock arithmetic." It leads to the construction of the integers modulo n. These are among the simplest examples of groups, as we shall see in Chapter 5. If n is a prime number, then the integers modulo n form a field. In Chapter 4, we will be looking at matrices with entries in these fields. As an application of congruences, we also discuss divisibility tests. In order to be able to solve linear congruences, we review greatest common divisors and the Euclidean algorithm.

Basic Properties

DEFINITION 1.1   *Fix a natural number n. The integers a and b are congruent modulo n or mod n, written*

$$a \equiv b \pmod{n},$$

if a − b is divisible by n.

For example,

$$23 \equiv 1 \pmod{11}$$
$$23 \equiv 2 \pmod{7}$$
$$23 \equiv -2 \pmod{25}.$$

If you measure time with a 12-hour clock, then you are calculating the hour modulo 12. For example, 5 hours after 9 o'clock is not 14 o'clock but 2 o'clock. We keep track of the days by reckoning modulo 7. If today is a Wednesday, then 10 days from today will be a Saturday. January 19 was a Wednesday in the year 2000. To determine what day of the week it was in 1998, we can calculate

$$2 \cdot 365 = 730 \equiv 2 \pmod{7}.$$

Therefore January 19 was a Friday in 1998. Calculating modulo n is very similar to calculating in the integers. First we note that congruence modulo n is an equivalence relation.

THEOREM 1.1

(i) $a \equiv a \pmod{n}$;

(ii) if $a \equiv b \pmod{n}$ then $b \equiv a \pmod{n}$;

(iii) if $a \equiv b \pmod{n}$ and $b \equiv c \pmod{n}$, then $a \equiv c \pmod{n}$.

It is easy to see why this is true. Clearly $a - a = 0$ is divisible by n. If $a - b$ is divisible by n then so is $b - a = -(a - b)$. And lastly, if $a - b$ and $b - c$ are divisible by n, then so is $a - c = (a - b) + (b - c)$.

Any integer a is congruent to a unique integer r, $0 \le r \le n - 1$. Simply divide a by n:

$$a = qn + r ,$$

for some q and r, $0 \le r < n$. Then $a \equiv r \pmod{n}$. From this you see that a is also congruent to a unique integer between 1 and n, or between -57 and $n - 58$. Addition and multiplication make sense modulo n:

THEOREM 1.2

(i) if $a \equiv b \pmod{n}$ and $c \equiv d \pmod{n}$, then $a + c \equiv b + d \pmod{n}$;

(ii) if $a \equiv b \pmod{n}$ and $c \equiv d \pmod{n}$, then $ac \equiv bd \pmod{n}$;

PROOF   Since $b - a$ and $d - c$ are divisible by n, then so are $(b + d) - (a + c) = (b - a) + (d - c)$ and $bd - ac = bd - bc + bc - ac = b(d - c) + c(b - a)$.   ∎

Divisibility Tests

With these simple properties we can establish some divisibility tests for natural numbers. Let us begin by deducing the obvious tests for divisibility by 2 and 5 using congruences. Suppose a is a natural number given in decimal form by $a = a_k \cdots a_1 a_0$, in other words

$$a = a_k 10^k + \cdots + a_1 10 + a_0$$

with $0 \le a_j < 10$ for all j. Then since $10^j \equiv 0 \pmod 2$ for any j,

$$a \equiv a_0 \pmod 2 .$$

So a is even if and only if its last digit a_0 is even. Similarly $10^j \equiv 0 \pmod 5$ so that

$$a \equiv a_0 \pmod 5 .$$

Thus a is divisible by 5 if and only if a_0 is, which is the case precisely when a_0 is 0 or 5. Next let us look at divisibility by 3 and 9.

TEST 1.1
Divisibility by 3 or 9

   (i) *A natural number a is divisible by 3 if and only if the sum of its digits is divisible by 3.*

   (ii) *A natural number a is divisible by 9 if and only if the sum of its digits is divisible by 9.*

PROOF   We have for $k > 0$,

$$x^k - 1 = (x - 1)(x^{k-1} + \cdots + x + 1) .$$

Taking $x = 10$, we see that $10^k - 1$ is divisible by 9 and in particular by 3. So

$$10^k \equiv 1 \pmod 9 \quad \text{and} \quad 10^k \equiv 1 \pmod 3 .$$

Therefore if

$$a = a_k 10^k + \cdots + a_1 10 + a_0 ,$$

then

$$a \equiv a_k + \cdots + a_1 + a_0 \pmod 9 \quad \text{and} \quad a \equiv a_k + \cdots + a_1 + a_0 \pmod 3 .$$

So a is divisible by 9, respectively 3, if and only if the sum of its digits is divisible by 9, respectively 3.   ∎

There is a test for divisibility by 11 which is similar. It is discussed in Exercise 1.5. The tests for divisibility by 7 and 13 are more subtle. Here is the test for 7. The test for 13 is in Exercise 1.6.

TEST 1.2
Divisibility by 7

   Let a be a natural number. Write $a = 10b + a_0$, where $0 \le a_0 < 10$. Then a is divisible by 7 if and only if $b - 2a_0$ is divisible by 7.

PROOF   We have

$$10b + a_0 \equiv 0 \quad (\text{mod } 7)$$

if and only if

$$10b + a_0 \equiv 21a_0 \quad (\text{mod } 7)$$

since $21a_0 \equiv 0 \ (\text{mod } 7)$. Equivalently,

$$10b - 20a_0 \equiv 0 \quad (\text{mod } 7),$$

i.e., 7 divides $10b - 20a_0 = 10(b - 2a_0)$. Since 10 is not divisible by 7, this holds if and only if 7 divides $b - 2a_0$. In other words,

$$b - 2a_0 \equiv 0 \quad (\text{mod } 7). \quad \blacksquare$$

For example,

$$426537183 \equiv 39 \equiv 0 \quad (\text{mod } 3), \quad \text{but} \quad 426537183 \equiv 39 \equiv 3 \quad (\text{mod } 9).$$

So 426537183 is divisible by 3 but not by 9. And

$$98 = 9 \cdot 10 + 8 \equiv 0 \quad (\text{mod } 7) \quad \text{since} \quad 9 - 2 \cdot 8 = -7 \equiv 0 \quad (\text{mod } 7).$$

Here is a table summarizing all the tests mentioned for a natural number a. In decimal form, a is given by $a_k 10^k + \cdots + a_1 10 + a_0 = 10b + a_0$.

n	test
2	a_0 even
5	$a_0 = 0$ or 5
3	$a_k + \cdots + a_1 + a_0$ divisible by 3
7	$b - 2a_0$ divisible by 7
9	$a_k + \cdots + a_1 + a_0$ divisible by 9
11	$a_k - \cdots + (-1)^{k-1}a_1 + (-1)^k a_0$ divisible by 11
13	$b + 4a_0$ divisible by 13

There is another divisibility question with a simple answer. When does a natural number n divide $10^k - 1$ for some $k > 0$? Not surprisingly this is related to the decimal expansion of $1/n$. First remember that every rational number m/n has a repeating decimal expansion (see [1], § 6.1). This expansion is finite if and only if the prime factors of the denominator n are 2 and 5. Otherwise it is infinite. If n divides $10^k - 1$, then the expansion of $1/n$ is of a special form. Suppose that

$$na = 10^k - 1, \quad a \in \mathbb{N}.$$

Write out the decimal expansion of a:

$$a = a_1 10^{k-1} + \cdots + a_{k-1} 10 + a_k.$$

So

$$a_1 10^{k-1} + \cdots + a_{k-1} 10 + a_k = \frac{10^k}{n} - \frac{1}{n} \, .$$

Divide this equation by 10^k:

$$0.a_1 \ldots a_{k-1} a_k = \frac{1}{n} - \frac{10^{-k}}{n} \, .$$

Divide again by 10^k:

$$0.\underbrace{0 \ldots 0}_{k} a_1 \ldots a_k = \frac{10^{-k}}{n} - \frac{10^{-2k}}{n} \, .$$

Continuing in this way, one gets

$$0.\underbrace{0 \ldots \ldots 0}_{ik} a_1 \ldots a_k = \frac{10^{-ik}}{n} - \frac{10^{-(i+1)k}}{n} \, ,$$

for any i. Now sum over i:

$$0.a_1 \ldots a_k a_1 \ldots a_k a_1 \ldots = \sum_{i=0}^{\infty} \left(\frac{10^{-ik}}{n} - \frac{10^{-(i+1)k}}{n} \right) = \left(\frac{1}{n} - \frac{10^{-k}}{n} \right) \sum_{i=0}^{\infty} 10^{-ik}$$

The sums converge because the series on the right is a geometric series. The sum in the middle telescopes, leaving $1/n$, and the left-hand side is a repeating decimal fraction. So

$$\frac{1}{n} = 0.a_1 \ldots a_k a_1 \ldots a_k a_1 \ldots \, . \tag{1.1}$$

Conversely, it is easy to show that if $1/n$ has a decimal expansion of this form, then n divides $10^k - 1$. The shortest such sequence of numbers $a_1, \ldots, a_k$ is called the *period* of $1/n$ and k the *length* of the period. If we have any other expansion for $1/n$,

$$\frac{1}{n} = 0.b_1 \ldots b_l b_1 \ldots b_l b_1 \ldots \, ,$$

for some $b_1, \ldots, b_l$, then we see that l must be a multiple of the period. Therefore the answer to our original question is:

THEOREM 1.3

$$10^k - 1 \equiv 0 \pmod{n}$$

if and only if $1/n$ *has a decimal expansion of the form* (1.1) *and k is a multiple of the length of the period of* $1/n$.

Taking $n = 7$, we can calculate that

$$\frac{1}{7} = 0.142857142857\ldots\,.$$

So 1, 4, 2, 8, 5, 7 is the period of $1/7$, which has length 6. Then $10^6 - 1 = 999999$ is divisible by 7, and for no smaller k is $10^k - 1$ divisible by 7.

Common Divisors

Recall that d is a *common divisor* of two integers a and b (not both 0) if d divides a and d divides b. The *greatest common divisor* is the largest one and is written (a, b). Every common divisor of a and b divides the greatest common divisor. We can compute (a, b) by the *Euclidean algorithm*. We divide a by b with a remainder r. Then we divide b by r with a remainder r_1, and so on until we get a remainder 0.

$$
\begin{aligned}
a &= qb + r & 0 &\le r < b \\
b &= q_1 r + r_1 & 0 &\le r_1 < r \\
&\ \ \vdots & &\ \ \vdots \\
r_{i-1} &= q_{i+1} r_i + r_{i+1} & 0 &\le r_{i+1} < r_i \\
&\ \ \vdots & &\ \ \vdots \\
r_{n-2} &= q_n r_{n-1} + r_n & 0 &\le r_n < r_{n-1} \\
r_{n-1} &= q_{n+1} r_n &
\end{aligned}
$$

In Chapter 14 we shall see the same algorithm for polynomials. The reason that this algorithm computes (a, b) is the following:

LEMMA 1.1

Let u and v be integers, not both 0. Write

$$u = qv + r\,,$$

for some integers q and r with $0 \le r < |v|$. *Then*

$$(u, v) = (v, r)\,.$$

PROOF If d is a common divisor of u and v, then d divides $r = u - qv$ and is therefore a common divisor of v and r. Conversely, if d is a common divisor of v and r, then d divides $u = qv + r$ and is therefore a common divisor of u and v. So the greatest common divisor of the two pairs must be the same. ∎

Applying this to the list of divisions above we obtain

$$(r_{i-1}, r_i) = (r_i, r_{i+1})$$

for each $i < n$. Now the last equation says that $r_n \mid r_{n-1}$. This means that

$$(r_{n-1}, r_n) = r_n .$$

Therefore arguing by induction,

$$(r_{i-1}, r_i) = r_n$$

for all i, in particular

$$(a, b) = r_n .$$

The proof of the lemma also shows that any common divisor of a and b divides (a, b).

We can read more out of the list of equations. The first equation tells us that r is a linear combination of a and b. The second one, that r_1 is a linear combination of b and r, and therefore of a and b. The third, that r_2 is a linear combination of r_1 and r, and therefore of a and b. Continuing like this, we get that r_n is a linear combination of a and b. Thus there exist integers s and t such that

$$(a, b) = sa + tb .$$

Example 1.1

Take $a = 57$ and $b = 21$. Then

$$57 = 2 \cdot 21 + 15$$
$$21 = 15 + 6$$
$$15 = 2 \cdot 6 + 3$$
$$6 = 2 \cdot 3 .$$

Therefore

$$3 = (57, 21) .$$

Furthermore

$$15 = 57 - 2 \cdot 21$$
$$6 = 21 - 15 = -57 + 3 \cdot 21$$
$$3 = 15 - 2 \cdot 6 = 3 \cdot 57 - 8 \cdot 21 .$$

So we can take $s = 3$ and $t = -8$. ☐

Another way to look at the Euclidean algorithm is as an algorithm that expresses the rational number a/b as a continued fraction. We have

$$\frac{a}{b} = q + \frac{r}{b} = q + \frac{1}{\left(\dfrac{b}{r}\right)}.$$

But

$$\frac{b}{r} = q_1 + \frac{r_1}{r},$$

so

$$\frac{a}{b} = q + \frac{1}{q_1 + \dfrac{r_1}{r}} = q + \frac{1}{q_1 + \dfrac{1}{\left(\dfrac{r}{r_1}\right)}}.$$

Continuing like this, our list of equations gives us the continued fraction

$$\frac{a}{b} = q + \cfrac{1}{q_1 + \cfrac{1}{\ddots \cfrac{1}{q_n + \cfrac{1}{q_{n+1}}}}}.$$

In our example, we find that

$$\frac{57}{21} = 2 + \cfrac{1}{1 + \cfrac{1}{2 + \cfrac{1}{2}}}.$$

For more about the origins of the Euclidean algorithm and its connection with the rational approximation of real numbers, see [2]. To read more about continued fractions in number theory, see [1].

Returning to the greatest common divisor, we say that a and b are *relatively prime* if $(a, b) = 1$, that is, if they have no common divisors except ± 1. Thus, if a and b are relatively prime, there exist integers s and t such that

$$1 = sa + tb.$$

For example, 16 and 35 are relatively prime and

$$1 = -5 \cdot 35 + 11 \cdot 16.$$

Related to the greatest common divisor of two integers is their least common multiple. The least common multiple m of $a, b \in \mathbb{Z}$ is the smallest common multiple of a and b, i.e., the smallest natural number that is divisible by both a and b. We shall write

$$m = \text{lcm}(a, b) .$$

It is not hard to see that m divides every common multiple of a and b, and that

$$\frac{ab}{(a, b)} = \text{lcm}(a, b) .$$

In Example 1.1 above,

$$\text{lcm}(57, 21) = 57 \cdot 21/3 = 399 .$$

Solving Congruences

It is very useful to be able to solve linear congruences, just the way you solve linear equations.

THEOREM 1.4
If $(a, n) = 1$, then the congruence

$$ax \equiv b \pmod{n}$$

has a solution which is unique modulo n.

PROOF Write $1 = as + nt$ for some integers s and t. Then $b = bas + bnt$. Thus

$$b \equiv abs \pmod{n}$$

and $x = bs$ is a solution of the congruence. If we have two solutions x and y, then $ax \equiv ay \pmod{n}$ i.e., n divides $ax - ay = a(x - y)$. Since a and n are relatively prime, this means that n divides $x - y$. Thus $x \equiv y \pmod{n}$. ∎

Example 1.2
Let us solve the congruence

$$24x \equiv 23 \pmod{31} .$$

First we use the Euclidean algorithm to find integers s and t such that $24s + 31t = 1$. One such pair is $s = -9$ and $t = 7$. Then

$$23 = 24(23s) + 31(23t)$$

so that
$$24(23s) \equiv 23 \quad (\mathrm{mod}\ 31)\ .$$

Thus $x = -207 \equiv 10$ (mod 31) is a solution of the congruence. We could also just compute multiples of 24 modulo 31. $\square$

In particular the congruence $ax \equiv 1$ (mod n) has a solution, unique modulo n, if $(a, n) = 1$. For example, if $a = 7$ and $n = 36$, then $5 \cdot 7 \equiv -1$ (mod 36). So $-5 \equiv 31$ (mod 36) is a solution of $7x \equiv 1$ (mod 36).

REMARK 1.1   If n is prime, then all a with $a \not\equiv 0$ (mod n) are relatively prime to n and have "multiplicative inverses" mod n.   ∎

Notice however that
$$9x \equiv 1 \quad (\mathrm{mod}\ 36)$$

does not have a solution. For if such an x did exist we could multiply the congruence by 4 and get
$$0 \equiv 4 \cdot 9x \equiv 4 \quad (\mathrm{mod}\ 36)\ .$$

But $0 - 4$ is not divisible by 36. In fact it is true, in general, that if a and n are not relatively prime, then
$$ax \equiv 1 \quad (\mathrm{mod}\ n)$$

does not have a solution. Suppose that $a = a'd$ and $n = n'd$ with $d > 1$. Then $n'a = n'a'd \equiv 0$ (mod n) so that

$$0 \equiv n'ax \equiv n' \not\equiv 0 \quad (\mathrm{mod}\ n)\ .$$

Therefore no such x can exist.

In general we let $\varphi(n)$ denote the number of integers a, $0 < a \le n$, which are relatively prime to n. For example if p is prime then $\varphi(p) = p - 1$.

Later we will also need to solve simultaneous congruences. The result that tells us this is possible is called the *Chinese Remainder Theorem*.

THEOREM 1.5
If n_1 and n_2 are relatively prime, then the two congruences

$$x \equiv b_1 \quad (\mathrm{mod}\ n_1) \quad \text{and} \quad x \equiv b_2 \quad (\mathrm{mod}\ n_2)$$

have a common solution, which is unique modulo $n_1 n_2$.

PROOF   A solution of the first congruence has the form $b_1 + sn_1$. For this to be a solution of the second congruence, we must have that

$$b_1 + sn_1 \equiv b_2 \quad (\mathrm{mod}\ n_2) \quad \text{or equivalently} \quad n_1 s \equiv b_2 - b_1 \quad (\mathrm{mod}\ n_2)\ .$$

Since $(n_1, n_2) = 1$, the previous theorem assures us that such an s exists. If x and y are two solutions, then $x \equiv y \pmod{n_1}$ and $x \equiv y \pmod{n_2}$, i.e., $x - y$ is divisible by n_1 and by n_2. As n_1 and n_2 are relatively prime, $x - y$ must be divisible by $n_1 n_2$. Thus $x \equiv y \pmod{n_1 n_2}$. ∎

Example 1.3
Solve

$$x \equiv 14 \pmod{24}$$
$$x \equiv 6 \pmod{31}.$$

Solutions of the first congruence are of the form $14 + 24s$. So we must solve

$$14 + 24s \equiv 6 \pmod{31}$$

or equivalently

$$24s \equiv 23 \pmod{31}.$$

This is the congruence we solved in Example 1.2. We saw that $s = 10$ is a solution. Therefore $x = 254$ is a solution of the two congruences we began with. ☐

The Integers Modulo n

As pointed out in Theorem 1.1, congruence modulo n is an equivalence relation. The equivalence classes are called congruence classes. The congruence class of an integer a is

$$\bar{a} := a + n\mathbb{Z} := \{a + sn \mid s \in \mathbb{Z}\}.$$

The set of all congruence classes is denoted by $\mathbb{Z}/n\mathbb{Z}$ and called the set of *integers modulo n*. Since every integer a is congruent to a unique r satisfying $0 \leq r < n$, $\bar{a}$ contains a unique r, $0 \leq r < n$. Thus there is a one-to-one correspondence between $\mathbb{Z}/n\mathbb{Z}$ and $\{0, 1, \dots, n-1\}$. For example, in $\mathbb{Z}/2\mathbb{Z}$ there are two congruence classes:

$$0 + 2\mathbb{Z} = \{2s \mid s \in \mathbb{Z}\},$$

the even integers, and

$$1 + 2\mathbb{Z} = \{1 + 2s \mid s \in \mathbb{Z}\},$$

the odd integers. We can define addition on $\mathbb{Z}/n\mathbb{Z}$ by

$$\bar{a} + \bar{b} = \overline{a + b}.$$

Because of Theorem 1.1, this makes sense. You can think of this as adding two natural numbers a and b in $\{0, 1, \dots, n-1\}$ modulo n, i.e., their sum is the remainder after division of $a + b$ by n. This addition in $\mathbb{Z}/n\mathbb{Z}$ is associative and commutative:

$$(\bar{a} + \bar{b}) + \bar{c} = \bar{a} + (\bar{b} + \bar{c})$$
$$\bar{a} + \bar{b} = \bar{b} + \bar{a}.$$

And

$$\bar{a} + \bar{0} = \bar{0}$$
$$\bar{a} + \overline{(-a)} = \bar{0}.$$

For example, in $\mathbb{Z}/10\mathbb{Z}$,

$$\bar{5} + \bar{7} = \bar{2}, \ \bar{4} + \bar{6} = \bar{0}.$$

Multiplication can also be defined on $\mathbb{Z}/n\mathbb{Z}$:

$$\bar{a} \cdot \bar{b} := \overline{ab}.$$

As with addition, we can think of this as just multiplication modulo n for two numbers from $\{0, 1, \ldots, n - 1\}$. It too is associative and commutative, and $\bar{1}$ is the identity element. If $(a, n) = 1$, then as pointed out in Remark 1.1, $\bar{a}$ has a multiplicative inverse. For example, we checked that in $\mathbb{Z}/36\mathbb{Z}$ the multiplicative inverse of $\bar{7}$ is $\overline{31}$. We also saw that because $\bar{4} \cdot \bar{9} = \bar{0}$, $\bar{9}$ has no multiplicative inverse.

We have seen that $\mathbb{Z}/n\mathbb{Z}$, at least when n is prime, has many formal properties in common with the set of real numbers $\mathbb{R}$ and complex numbers $\mathbb{C}$. We can collect these properties in a formal definition.

DEFINITION 1.2   *A field F is a set with two binary operations, called "addition" and "multiplication," written $+$ and $\cdot$, respectively, with the following properties (a binary operation on F is just a mapping $F \times F \to F$):*

(i) *Addition and multiplication are both associative and commutative;*

(ii) *For $a, b, c \in F$, $a(b + c) = ab + ac$;*

(iii) *There exist distinct elements $0, 1 \in F$ such that for any $a \in F$,*

$$a + 0 = a$$
$$a \cdot 1 = a.$$

(iv) *For any $a \in F$ there exists a unique element, written $-a$, such that*

$$a + (-a) = 0,$$

and for any $a \in F$, $a \neq 0$, there exists a unique element, written a^{-1}, such that

$$a \cdot a^{-1} = 1.$$

These are the formal properties satisfied by addition and multiplication in $\mathbb{R}$ and $\mathbb{C}$, as well as in $\mathbb{Z}/p\mathbb{Z}$ for p prime. We introduce the notation $\mathbb{F}_p$ for $\mathbb{Z}/p\mathbb{Z}$. As pointed out above, if n is not prime, then not all nonzero elements in $\mathbb{Z}/n\mathbb{Z}$ have multiplicative inverses. In these cases $\mathbb{Z}/n\mathbb{Z}$ is not a field.

Introduction to Software

The main purpose of this section is to give you a chance to practice using *Mathematica*. It has several functions which are relevant to this section. For making computations in the integers modulo *n*, there is a built-in function Mod. Thus

    In[1] := Mod[25+87,13]

    Out[1] = 8

and

    In[2] := Mod[2^12,7]

    Out[2] = 1

A more efficient way of computing powers mod *n* is to use the function PowerMod:

    In[3] := PowerMod[2,12,7]

    Out[3] = 1

For any real number *a* the function N[a, m] will compute the first m digits of a. For example,

    In[4] := N[1/7, 20]

    Out[4] = 0.14285714285714285714

So we see that the period of 1/7 is 1, 4, 2, 8, 5, 7, as discussed above. Similarly

    In[5] := N[1/19, 40]

    Out[5] = 0.0526315789473684210526315789473684210526

Thus the period of 1/19 is 0, 5, 2, 6, 3, 1, 5, 7, 8, 9, 4, 7, 3, 6, 8, 4, 2, 1 , which has length 18. And

    In[6] := Mod[10^18 - 1, 19]

    Out[6] = 0

but for smaller k, $10^k - 1 \not\equiv 0$ (mod 19). We can do these calculations for all the primes less than say, 30, and tabulate the results. They confirm Theorem 1.3.

prime	3	7	11	13	17	19	23	29
length of period	2	6	2	6	16	18	22	28
smallest k	2	6	2	6	16	18	22	28 .

Interestingly

$$1001 = 10^3 + 1 = 7 \cdot 11 \cdot 13 \,,$$

so that

$$10^3 + 1 \equiv 0 \pmod 7 \,.$$

Similarly,

```
In[7]:= Mod[10^9+1, 19]

Out[7]= 0
```

What do you think is going on? Experiment a bit and try to make a conjecture.

   Given a pair of integers a and b, the function ExtendedGCD will compute (a, b) and integers s and t such that $(a, b) = as + bt$. For example,

```
In[8]:= ExtendedGCD[57, 21]

Out[8]= {3,{3,-8}}
```

This means that $3 = 3 \cdot 57 - 8 \cdot 21$. If we want to find the multiplicative inverse of $\overline{105}$ in $\mathbb{F}_{197}$, we can enter

```
In[9]:= ExtendedGCD[105, 197]

Out[9]= {1,{-15,8}}
```

So $1 = -15 \cdot 105 + 8 \cdot 197$ and $-\overline{15} = \overline{182}$ is the multiplicative inverse of $\overline{105}$.

   In the package NumberTheory'NumberTheoryFunctions' there is a function called ChineseRemainderTheorem, which will solve two such congruences simultaneously. In fact it will solve a system:

$$x \equiv b_1 \pmod{n_1}$$

$$\vdots$$

$$x \equiv b_r \pmod{n_r} \,.$$

To use it you first load the package:

```
In[10]:= « NumberTheory'NumberTheoryFunctions'
```

Then you simply enter

```
In[11]:= ChineseRemainderTheorem[{b₁, ... , bᵣ},
           {n₁, ... , nᵣ}]
```

So in our example:

```
In[12]:= ChineseRemainderTheorem[{14, 6},{24, 31}]

Out[12]= 254
```

Exercises

1.1 On what day of the week will January 19 fall in the year 2003?

1.2 Solve the following congruences:

   (a) $5x \equiv 6 \pmod{7}$

   (b) $5x \equiv 6 \pmod{8}$

   (c) $3x \equiv 2 \pmod{24}$

   (d) $14x \equiv 5 \pmod{45}$.

1.3 Solve the following systems of congruences:

   (a) $5x \equiv 6 \pmod{7}$ and $5x \equiv 6 \pmod{8}$

   (b) $14x \equiv 5 \pmod{45}$ and $4x \equiv 5 \pmod{23}$.

   Check whether the *Mathematica* function ChineseRemainderTheorem gives the same answers as you get by hand.

1.4 • For p prime, $0 < k < p$, show that

   (a)
$$\binom{p}{k} \equiv 0 \pmod{p} ;$$

   (b)
$$(x + y)^p \equiv x^p + y^p \pmod{p} .$$

1.5 Prove that a natural number a is divisible by 11 if and only if the alternating sum of its digits is divisible by 11.

1.6 Show that $10m + n, 0 \le n < 10$, is divisible by 13 if and only if $m + 4n$ is divisible by 13.

1.7 Prove that a rational number m/n, where m and n are integers, $n \neq 0$, has a finite decimal expansion if and only if the prime factors of n are 2 and 5.

1.8 Let n be a natural number. Suppose that $1/n$ has a decimal expansion of the form $0.a_1 \ldots a_k a_1 \ldots a_k a_1 \ldots$. Prove that $n \mid (10^k - 1)$.

1.9 For each prime $p < 30$, find the smallest $k \in \mathbb{N}$ such that

$$10^k + 1 \equiv 0 \pmod{p},$$

and compare it with the length of the period of $1/p$.

1.10 Let p be a prime number, and let k be the length of the period of $1/p$. Suppose that $k = 2l$ is even. Prove that $10^l + 1$ is divisible by p, and that $10^m + 1$ is not divisible by p for any $m < l$. (Suggestion: write $10^{2l} - 1 = (10^l - 1)(10^l + 1)$). Does this hold true if p is a composite number?

1.11 • Make a table of $\varphi(n)$ for $n \leq 20$. For p prime, $r \geq 1$, calculate $\varphi(p^r)$.

1.12 Suppose that you have a bucket that holds 57 cups, and one that holds 21 cups. How could you use them to measure out 3 cups of water?

1.13 Given integers a and b, which are relatively prime, suppose that

$$sa - tb = 1,$$

for some integers s and t. Suppose that

$$s'a - t'b = 1,$$

as well, for some integers s' and t'. Prove that there then exists an integer k such that
$$s' = s + kb \quad \text{and} \quad t' = t + ka.$$

1.14 Let a and b be two nonzero integers.

(a) Show that $\text{lcm}(a, b)$ divides every common multiple of a and b.

(b) Suppose that $(a, b) = 1$. Prove that then $\text{lcm}(a, b) = ab$.

(c) Prove that in general

$$\text{lcm}(a, b) = \frac{ab}{(a, b)}.$$

1.15 Write out addition and multiplication tables for $\mathbb{Z}/6\mathbb{Z}$. Which elements have multiplicative inverses?

1.16 Write out the multiplication table for $\mathbb{Z}/7\mathbb{Z}$. Make a list of the multiplicative inverses of the nonzero elements.

1.17 How many elements of $\mathbb{Z}/7\mathbb{Z}$ and $\mathbb{Z}/11\mathbb{Z}$ are squares? How many elements of $\mathbb{Z}/p\mathbb{Z}$, where p is prime, are squares? Make a conjecture and then try to prove it.

1.18 Suppose that $n \in \mathbb{N}$ is a composite number.

    (a) Show that there exist $a, b \in \mathbb{Z}$ with $\bar{a}, \bar{b} \neq 0 \in \mathbb{Z}/n\mathbb{Z}$, but $\bar{a}\bar{b} = 0$.

    (b) Prove that $\mathbb{Z}/n\mathbb{Z}$ is not a field.

1.19 • Let
$$\mathbb{Q}\left(\sqrt{2}\right) = \left\{a + b\sqrt{2} \mid a, b \in \mathbb{Q}\right\} \subset \mathbb{R}.$$

Verify that $\mathbb{Q}(\sqrt{2})$ is a field.

1.20 Let
$$\mathbf{C} = \left\{\begin{pmatrix} a & b \\ -b & a \end{pmatrix} \middle| a, b \in \mathbb{R}\right\}$$

Show that $\mathbf{C}$ is a field under the operations of matrix addition and multiplication.

1.21 • Let
$$\mathbb{F}_{p^2} = \left\{\begin{pmatrix} a & b \\ br & a \end{pmatrix} \middle| a, b \in \mathbb{F}_p\right\}$$

where $r \in \mathbb{F}_p$ is not a square. Show that under the operations of matrix addition and matrix multiplication, $\mathbb{F}_{p^2}$ is a field with p^2 elements. Suggestion: to prove that a nonzero matrix has a multiplicative inverse, use the fact that the congruence $x^2 - r \equiv 0 \pmod{p}$ has no solution. What happens if r is a square in $\mathbb{F}_p$?

Chapter 2

Permutations

Permutations as Mappings

In the next chapter, we will begin looking at groups by studying permutation groups. To do this we must first establish the properties of permutations that we shall need there. A permutation of a set X is a rearrangement of the elements of X. More precisely,

DEFINITION 2.1   *A permutation of a set X is a bijective mapping of X to itself.*

A bijective mapping is a mapping that is one-to-one and onto. We are mainly interested in permutations of finite sets, in particular of the sets $\{1, 2, \ldots, n\}$ where n is a natural number. A convenient way of writing such a permutation α is the following. Write the numbers $1, 2, \ldots, n$ in a row and write their images under α in a row beneath:

$$\begin{pmatrix} 1 & 2 & \ldots & n \\ \alpha(1) & \alpha(2) & \ldots & \alpha(n) \end{pmatrix}.$$

For example, the permutation α of $\{1, 2, 3, 4, 5\}$ with $\alpha(1) = 3$, $\alpha(2) = 1$, $\alpha(3) = 5$, $\alpha(4) = 2$, and $\alpha(5) = 4$ is written

$$\alpha = \begin{pmatrix} 1 & 2 & 3 & 4 & 5 \\ 3 & 1 & 5 & 2 & 4 \end{pmatrix}.$$

This notation is usually called a *mapping notation*.

We denote by S_X, the set of all permutations of a set X, and by S_n, the set of permutations of $\{1, 2, \ldots, n\}$. It is easy to count the number of permutations in S_n. A permutation can map 1 to any of $1, 2, \ldots, n$. Having chosen the image of 1, 2 can be mapped to any of the remaining $n - 1$ numbers. Having chosen the images of 1 and 2, 3 can be mapped to any of the remaining $n - 2$ numbers. And so on, until the images of $1, \ldots, n - 1$ have been chosen. The remaining number must be the image of n. So there are $n(n - 1)(n - 2) \cdots 1$ choices. Thus S_n has $n!$ elements.

If we are given two permutations α and β of a set X, then the composition $\alpha \circ \beta$ is also one-to-one and onto. So it too is a permutation of the elements of X. For example, if α is the element of S_5 above, and

$$\beta = \begin{pmatrix} 1\ 2\ 3\ 4\ 5 \\ 2\ 3\ 5\ 1\ 4 \end{pmatrix},$$

then

$$\alpha \circ \beta = \begin{pmatrix} 1\ 2\ 3\ 4\ 5 \\ 1\ 5\ 4\ 3\ 2 \end{pmatrix}.$$

The entries in the second row are just $(\alpha \circ \beta)(1), \ldots , (\alpha \circ \beta)(5)$. Notice that

$$\beta \circ \alpha = \begin{pmatrix} 1\ 2\ 3\ 4\ 5 \\ 5\ 2\ 4\ 3\ 1 \end{pmatrix} \neq \alpha \circ \beta .$$

The order in which you compose permutations matters.

Since a permutation is bijective, it has an inverse, which is also a bijective mapping. The inverse of the element $\beta \in S_5$ above is

$$\beta^{-1} = \begin{pmatrix} 1\ 2\ 3\ 4\ 5 \\ 4\ 1\ 2\ 5\ 3 \end{pmatrix}.$$

To compute the entries in the second row, find the number that β maps to 1, to 2, etc. You can quickly check that

$$\beta^{-1} \circ \beta = \beta \circ \beta^{-1} = \begin{pmatrix} 1\ 2\ 3\ 4\ 5 \\ 1\ 2\ 3\ 4\ 5 \end{pmatrix},$$

which is the identity permutation.

From now on, for convenience, we are going to write the composition of two permutations α and β as a "product:"

$$\alpha\beta := \alpha \circ \beta .$$

So to calculate $(\alpha\beta)(i)$, we first apply β to i and then apply α to the result. In other words, we read products "from right to left".

Cycles

Let

$$\alpha = \begin{pmatrix} 1\ 2\ 3\ 4\ 5 \\ 3\ 2\ 4\ 1\ 5 \end{pmatrix} \in S_5 .$$

So $\alpha(1) = 3$, $\alpha(3) = 4$, $\alpha(4) = 1$ and α fixes 2 and 5. We say that α permutes 1, 3, and 4 cyclically and that α is a cycle, or more precisely, a 3-cycle. In general, an element

$\alpha \in S_n$ is an r-cycle, where $r \leq n$, if there is a sequence $i_1, i_2, \ldots, i_r \in \{1, \ldots, n\}$ of distinct numbers, such that

$$\alpha(i_1) = i_2, \ \alpha(i_2) = i_3, \ \ldots, \ \alpha(i_{r-1}) = i_r, \ \alpha(i_r) = i_1 ,$$

and α fixes all other elements of $\{1, \ldots, n\}$.

Cycles are particularly simple. We shall show that any permutation can be written as a product of cycles, in fact, there is even a simple algorithm that does this. Let us first carry it out in an example. Take

$$\alpha = \begin{pmatrix} 1\,2\,3\,4\,5\,6\,7\,8 \\ 2\,4\,5\,1\,3\,8\,6\,7 \end{pmatrix} .$$

We begin by looking at $\alpha(1)$, $\alpha^2(1)$, $\ldots$. We have

$$\alpha(1) = 2, \ \alpha(2) = 4, \ \alpha(4) = 1 .$$

As our first cycle α_1, then we take the 3-cycle

$$\alpha_1 = \begin{pmatrix} 1\,2\,3\,4\,5\,6\,7\,8 \\ 2\,4\,3\,1\,5\,6\,7\,8 \end{pmatrix} .$$

The smallest number that does not occur in this cycle is 3. We see that

$$\alpha(3) = 5, \ \alpha(5) = 3 .$$

So take as our second cycle, the 2-cycle

$$\alpha_2 = \begin{pmatrix} 1\,2\,3\,4\,5\,6\,7\,8 \\ 1\,2\,5\,4\,3\,6\,7\,8 \end{pmatrix} .$$

The smallest number that is left fixed by α_1 and α_2 is 6. Now

$$\alpha(6) = 8, \ \alpha(8) = 7, \ \alpha(7) = 6 .$$

So let

$$\alpha_3 = \begin{pmatrix} 1\,2\,3\,4\,5\,6\,7\,8 \\ 1\,2\,3\,4\,5\,8\,6\,7 \end{pmatrix} ,$$

The numbers $\{1, 2, \ldots, 8\}$ are all accounted for now, and we see that

$$\alpha = \alpha_3 \alpha_2 \alpha_1 .$$

The cycles are even disjoint, that is, no two have a number in common. This works in general:

THEOREM 2.1
Any permutation is a product of disjoint cycles.

To see this, let $\alpha \in S_n$. Consider $\alpha(1)$, $\alpha^2(1)$, ... At some point this sequence will begin to repeat itself. Suppose that $\alpha^t(1) = \alpha^s(1)$ where $s < t$. Then $\alpha^{t-s}(1) = 1$. Pick the smallest $r_1 > 0$ such that $\alpha^{r_1}(1) = 1$. Let α_1 be the r_1-cycle given by the sequence

$$1, \ \alpha(1), \ \alpha^2(1), \ \dots , \ \alpha^{r_1-1}(1) \ .$$

Now pick the smallest number $i_2 \neq \alpha^i(1)$ for any i. Consider $\alpha(i_2)$, $\alpha^2(i_2)$, ... Again pick the smallest r_2 such that $\alpha^{r_2}(i_2) = i_2$ and let α_2 be the r_2-cycle given by the sequence

$$i_2, \ \alpha(i_2), \ \alpha^2(i_2), \ \dots , \ \alpha^{r_2-1}(i_2) \ .$$

Continuing this way we find cycles $\alpha_1, \alpha_2, \dots , \alpha_k$ such that

$$\alpha \ = \ \alpha_k \cdots \alpha_2 \alpha_1 \ .$$

And these cycles are disjoint from one another.

At this point it is convenient to introduce *cycle notation*. If $\alpha \in S_n$ is an r-cycle, then there exist $i_1, \dots , i_r \in \{1, \dots , n\}$ distinct from one another such that

$$\alpha(i_k) \ = \ i_{k+1} \ , \ \text{for } 1 \leq k < r \ , \ \alpha(i_r) \ = \ i_1, \ \alpha(j) \ = \ j \text{ otherwise} \ .$$

Then we write

$$\alpha \ = \ (i_1 \ i_2 \ \cdots \ i_r) \ .$$

So in the example above,

$$\alpha_1 \ = \ (1\,2\,4) \ , \ \alpha_2 \ = \ (3\,5) \ , \ \alpha_3 \ = \ (6\,8\,7) \ .$$

And in cycle notation, we write

$$\alpha \ = \ (1\,2\,4)(3\,5)(6\,8\,7) \ .$$

We do not write out 1-cycles, except with the identity permutation, which is written (1) .

The order in which you write the cycles in a product of disjoint cycles does not matter for the following reason:

THEOREM 2.2
Disjoint cycles commute with each other.

To see this, suppose that $\alpha, \beta, \in S_n$ are disjoint cycles given by

$$\alpha \ = \ (i_1 \ \cdots \ i_r) \ , \ \beta \ = \ (j_1 \ \cdots \ j_s) \ .$$

Then

$$\begin{cases} \alpha(\beta(j)) = j = \beta(\alpha(j)) \ , & \text{for } j \notin \{i_1, \dots , i_r, j_1, \dots , j_s\} \ , \\ \alpha(\beta(i_k)) = i_{k+1} = \beta(\alpha(i_k)) \ , & \text{for } 1 \leq k \leq r \ , \\ \alpha(\beta(j_k)) = j_{k+1} = \beta(\alpha(j_k)) \ , & \text{for } 1 \leq k \leq s \ . \end{cases}$$

It is understood that $i_{r+1} := i_1$ and $j_{s+1} := j_1$.

A 2-cycle is called a transposition. Any cycle can be written as a product of transpositions. For example,

$$(1\,2\,3\,4) \;=\; (1\,4)(1\,3)(1\,2)$$

or equally well,

$$(1\,2\,3\,4) \;=\; (1\,2)(2\,4)(2\,3)\,.$$

In general, if $(i_1\ i_2\ \cdots\ i_r) \in S_n$, then

$$(i_1\ i_2\ \cdots\ i_r) \;=\; (i_1\ i_r)\cdots(i_1\ i_3)(i_1\ i_2)\,.$$

Combining this with the theorem above, we see that

THEOREM 2.3
Any permutation can be written as a product of transpositions.

For example

$$\begin{pmatrix} 1\,2\,3\,4\,5\,6\,7\,8 \\ 2\,4\,5\,1\,3\,8\,6\,7 \end{pmatrix} \;=\; (1\,2\,4)(3\,5)(6\,8\,7) \;=\; (1\,4)(1\,2)(3\,5)(6\,7)(6\,8)\,.$$

Actually this theorem is intuitively obvious. If you want to reorder a set of objects, you naturally do it by switching them in pairs.

Sign of a Permutation

In writing a permutation as a product of transpositions, the number of transpositions is not well-defined. For example,

$$(2\,3)(1\,2\,3) \;=\; (2\,3)(1\,3)(1\,2)$$

and

$$(2\,3)(1\,2\,3) \;=\; (1\,3)\,.$$

The parity of this number does turn out to be well-defined either: the permutation can be written as the product of an even number or an odd number of transpositions, but not both.

To see this, let A be a real $n \times n$ matrix and $\alpha \in S_n$. Let A_α denote the matrix obtained from A by permuting the rows according to α. So the first row of A_α is the $\alpha(1)$th row of A, the second row of A_α is the $\alpha(2)$th row of A, and so on. This is sometimes called a row operation on A. Recall that

$$A_\alpha \;=\; I_\alpha A\,,$$

where I is the identity matrix. In particular, taking $A = I_\beta$,

$$(I_\beta)_\alpha = I_\alpha I_\beta .$$

Now $(I_\beta)_\alpha = I_{\alpha\beta}$ (because we are reading products from right to left). So

$$I_{\alpha\beta} = I_\alpha I_\beta ,$$

and therefore
$$\det(I_{\alpha\beta}) = \det(I_\alpha)\det(I_\beta) .$$

If α is a transposition, then
$$\det(I_\alpha) = -1 .$$

(Interchanging two rows of a determinant changes its sign.) So if α is the product of r transpositions, then
$$\det(I_\alpha) = (-1)^r .$$

The left-hand side depends only on α, and the right-hand side only depends on the parity of r.

We define the *sign* of α, written sgn α by

$$\mathrm{sgn}\,\alpha := \det(I_\alpha) .$$

A permutation α is *even* if sgn α is 1, i.e., if α can be written as a product of an even number of transpositions, and *odd* if sgn α is -1. Thus the product of two even permutations is even, of two odd ones even, and of an even one and an odd one odd.

Exercises

2.1 For the permutation
$$\alpha = \begin{pmatrix} 1\ 2\ 3\ 4\ 5 \\ 3\ 1\ 5\ 2\ 4 \end{pmatrix} ,$$

compute α^2 and α^3. What is the smallest power of α, which is the identity?

2.2 • In S_3, let
$$\alpha = \begin{pmatrix} 1\ 2\ 3 \\ 2\ 3\ 1 \end{pmatrix} , \quad \beta = \begin{pmatrix} 1\ 2\ 3 \\ 2\ 1\ 3 \end{pmatrix} .$$

Compute $\alpha^2, \alpha^3, \beta^2, \alpha\beta, \alpha^2\beta$. Check that these, together with α and β, are the six elements of S_3. Verify that

$$\alpha^2 = \alpha^{-1} , \ \beta = \beta^{-1} , \ \alpha^2\beta = \beta\alpha .$$

2.3 • Suppose that α and β are permutations. Show that $(\alpha\beta)^{-1} = \beta^{-1}\alpha^{-1}$

2.4 How many 3-cycles are there in S_4? Write them out.

2.5 How many 3-cycles are there in S_n for any n? How many r-cycles are there in S_n for an arbitrary $r \leq n$?

2.6 Prove that if α is an r-cycle, then α^r is the identity permutation.

2.7 Two permutations α and β are said to be disjoint if $\alpha(i) \neq i$ implies that $\beta(i) = i$ and $\beta(j) \neq j$ implies that $\alpha(j) = j$. Prove that disjoint permutations commute with one another.

2.8 Write the following permutations as products of cycles:

(a)
$$\begin{pmatrix} 1\,2\,3\,4\,5\,6\,7\,8\,9 \\ 4\,6\,7\,1\,5\,2\,8\,3\,9 \end{pmatrix}$$

(b)
$$\begin{pmatrix} 1\,2\,3\,4\,5\,6\,7\,8\,9 \\ 6\,1\,7\,5\,4\,2\,8\,9\,3 \end{pmatrix}$$

2.9 Write the two permutations in the previous exercise as products of transpositions.

2.10 Show that the inverse of an even permutation is even, of an odd permutation odd.

Chapter 3

Permutation Groups

Definition

Suppose you have a square and number its vertices.

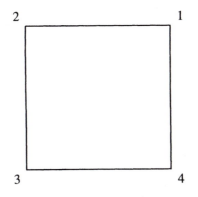

Each symmetry of the square permutes the vertices, and thus gives you an element of S_4. We can make a table showing the eight symmetries of the square and the corresponding permutations:

Symmetry	Permutation
rotation counterclockwise through $\pi/2$	(1 2 3 4)
rotation counterclockwise through π	(1 3)(2 4)
rotation counterclockwise through $3\pi/2$	(1 4 3 2)
identity map	(1)
reflection in diagonal through 1 and 3	(2 4)
reflection in diagonal through 2 and 4	(1 3)
reflection in vertical axis	(1 2)(3 4)
reflection in horizontal axis	(1 4)(2 3)

Let $D_4 \subset S_4$ denote the set of permutations in the right-hand column,

$$D_4 = \{(1\,2\,3\,4), (1\,3)(2\,4), (1\,4\,3\,2), (1), (2\,4), (1\,3), (1\,2)(3\,4), (1\,4)(2\,3)\}\,.$$

Now the set of all symmetries of the square has the following two properties:

(i) the composition of two symmetries is a symmetry;

(ii) the inverse of a symmetry is a symmetry.

It is easy to see that under the correspondence above that products map to products. So the set D_4 will have the same properties. You can also check this directly. Sets of permutations with these algebraic properties are called permutation groups. As we shall see they arise in many contexts.

DEFINITION 3.1   *A nonempty set of permutations $G \subset S_n$ is called a permutation group (of degree n) if for all $\alpha, \beta \in G$*

$$(i) \qquad \alpha\beta \in G$$
$$(ii) \qquad \alpha^{-1} \in G\,.$$

S_n itself is a permutation group, called the full permutation group (of degree n) or *symmetric group* (of degree n). Another example is

$$V' = \{(1\,2), (3\,4), (1\,2)(3\,4), (1)\} \subset S_4\,.$$

We see that

$$(1\,2)(3\,4) \cdot (1\,2) = (1\,2) \cdot (1\,2)(3\,4) = (3\,4)$$
$$(1\,2)(3\,4) \cdot (3\,4) = (3\,4) \cdot (1\,2)(3\,4) = (1\,2)$$
$$(3\,4) \cdot (1\,2) = (1\,2) \cdot (3\,4) = (1\,2)(3\,4)$$
$$(1\,2)^2 = (3\,4)^2 = \left((1\,2)(3\,4)\right)^2 = (1)$$

$$\left((1\,2)(3\,4)\right)^{-1} = (1\,2)(3\,4)\,, \quad (1\,2)^{-1} = (1\,2)\,, \quad (3\,4)^{-1} = (3\,4)\,.$$

So V' is a permutation group. With appropriate software it is easy to check whether a set of permutations is a permutation group. This will be explained below.

Let $A_n \subset S_n$ denote the set of all even permutations. In Chapter 2, it was pointed out that the product of two even permutations is even and that the inverse of an even permutation is also even. Therefore A_n is a permutation group, called the *alternating group* (of degree n). Of the six elements of S_3, three are even: the two 3-cycles and the identity. Thus

$$A_3 = \{(1), (1\,2\,3), (1\,3\,2)\}\,.$$

The number of elements in a permutation group G is called the *order* of G, written $|G|$. Thus $|S_n| = n!$, $|A_3| = 3$, and $|V'| = 4$.

Cyclic Groups

There is a very simple class of permutation groups. You construct them in the following way. Take any $\alpha \in S_n$, for some n. Consider the powers of α: $\alpha, \alpha^2, \ldots$. Since S_n is finite, at some point an element in this list will be repeated. Suppose that

$$\alpha^t = \alpha^s, \quad \text{for some } s < t .$$

Then multiplying both sides by α^{-s}, we see that

$$\alpha^{t-s} = (1) .$$

Let r be the smallest natural number such that

$$\alpha^r = (1) .$$

Set

$$G = \{(1), \alpha, \ldots, \alpha^{r-1}\} \subset S_n .$$

Now check that G is a permutation group: we have

$$(\alpha^i)^{-1} = \alpha^{-i} = \alpha^{r-i} \quad \text{for any } i, \ 1 \leq i < r$$

and

$$\alpha^i \alpha^j = \alpha^{i+j} = \alpha^k , \quad \text{where } i + j \equiv k \pmod{r}, \ 0 \leq k < r .$$

G is called the cyclic permutation group generated by α and will be denoted by $\langle \alpha \rangle$. It has order r. As you can imagine there is a connection between a cyclic permutation group of order r and the integers mod r. This will be made clear in Theorem 6.1.

Examples 3.1

(i) Take $\alpha = (1\,2)$. Since $(1\,2)^2 = (1)$, $\{(1), (1\,2)\}$ is a cyclic permutation group of order 2.

(ii) Take $\alpha = (1\,2\,3)$. Since $(1\,2\,3)^3 = (1)$, and $(1\,2\,3)^2 = (1\,3\,2)$, $\{(1), (1\,2\,3), (1\,3\,2)\}$ is a cyclic permutation group of order 3. In fact, this group is the alternating group of degree 3, A_3. ▯

These examples suggest that the cyclic permutation group generated by an r-cycle should have order r. Indeed, if

$$\alpha = (i_1\ i_2\ \cdots\ i_r) ,$$

then $\alpha^r = (1)$, but for $j < r$, $\alpha^j(i_1) = i_j \neq i_1$, so that $\alpha^j \neq (1)$. Therefore the cyclic permutation group generated by α does have order r.

DEFINITION 3.2   *The order of a permutation α, written $|\alpha|$, is the order of the cyclic permutation group $\langle \alpha \rangle$ generated by α, or equivalently, the smallest $r \in \mathbb{N}$ such that $\alpha^r = (1)$.*

REMARK 3.1   Suppose that $\alpha^s = 1$, for some $s \in \mathbb{N}$. Let $r = |\alpha|$. Write $s = qr + t$, for some $q, t \in \mathbb{Z}$ where $0 \leq t < r$. Then

$$(1) = \alpha^s = \alpha^{qr+t} = \alpha^t .$$

But then by the definition of $|\alpha|$ we must have $t = 0$. So r divides s.   ∎

Example 3.2
Suppose we write α as a product of disjoint cycles,

$$\alpha = \alpha_1 \alpha_2 \cdots \alpha_k$$

where α_i is an r_i-cycle. Suppose $\alpha^s = 1$, for some s. Since $\alpha_1, \alpha_2, \ldots, \alpha_k$ are disjoint cycles, this implies that

$$\alpha_i^s = (1)$$

for all i, $1 \leq i \leq k$. (Check this!) But then by the remark above, r_i divides s for all i. So s is a common multiple of these orders. Therefore the order of α is the least common multiple of $r_1, r_2, \ldots, r_k$.   ⬚

Generators

The group V' above is not cyclic. Each of its elements has order 2 except for the identity. There is no element of order 4. However, our calculation shows that if we begin with, say, $(1\,2)$ and $(3\,4)$, we can express the remaining two elements of V' in terms of $(1\,2)$ and $(3\,4)$. Similarly, in Exercise 2.2 we saw that every element of S_3 can be written in terms of $(1\,2)$ and $(1\,2\,3)$. We say that V' or S_3 is generated by $\{(1\,2)(3\,4), (1\,3)(2\,4)\}$, respectively $\{(1\,2), (1\,2\,3)\}$. In general a permutation group G is said to be generated by a subset $g \subset G$ if every element in G can be written as a product of elements of g.

Theorem 2.3 tells us that the set of all transpositions generates S_n. This is a relatively large set: there are $\binom{n}{2}$ transpositions in S_n. In fact, the set of $n - 1$ transpositions

$$g = \{(1\,2), (2\,3), \cdots, (n-1\,n)\}$$

will do. To see this we just need to prove that any transposition can be expressed in terms of transpositions in g. Because every permutation can be written in terms of transpositions, it follows then that every permutation can be written in terms of elements of g. Take a transposition $(i\ j)$, where $i < j$. We have

$$(i\ j) = (j-1\ j)(j-2\ j-1)\cdots(i\ i+1)\cdots(j-2\ j-1)(j-1\ j)\,.$$

But we can do even better than that. Just as in the case of S_3, we can generate S_n using just one transposition and one n-cycle. Let

$$\alpha = (1\ 2)\quad,\quad \beta = (1\ 2\ \cdots\ n)\,.$$

Then for $1 < i < n$,

$$(i\ i+1) = \beta^{i-1}\alpha\beta^{-i+1} = \beta^{i-1}\alpha\beta^{n-i+1}\,.$$

So every transposition in g and therefore every permutation can be written in terms of α and β and thus $\{\alpha, \beta\}$ generates S_n.

In practice, it is very clumsy to describe a permutation group by listing all its elements. In fact, for any permutation group of moderately large size it is impossible — try writing out all the elements of S_{10}! A more convenient way is to give a set of generators for it. Suppose we begin with a set of permutations $g \subset S_n$ for some n. What do we mean by *the permutation group G generated by g* ? In some sense, G is the set of all permutations that can be expressed in terms of elements of g. The following theorem makes this more precise.

THEOREM 3.1
Let G be the smallest permutation group containing g. Then

$$G = \bigcup_{i=0}^{\infty} g^i\,.$$

The set g^i is the set of all products of i elements of g, i.e.,

$$g^i = \{\alpha_1\alpha_2\cdots\alpha_i \mid \alpha_1, \alpha_2, \ldots, \alpha_i \in g\}\,.$$

PROOF Let H denote the right-hand side. If G is a permutation group containing g, then $g^i \subset G$ for all i. So H is contained in any such G. We just need to prove that H is a permutation group. Well, the product of any two elements in H lies in H. And if $\alpha \in g$, then as we saw in the previous section, $\alpha^{-1} \in g^i$ for some i. But for any $\alpha_1, \ldots, \alpha_i \in g$, $(\alpha_1 \cdots \alpha_i)^{-1} = \alpha_i^{-1} \cdots \alpha_1^{-1}$ (cf. Exercise 2.3). So the inverse of any element in H also lies in H. Thus H is a permutation group and therefore $G = H$. ∎

Now we can make our definition:

DEFINITION 3.3   *The permutation group generated by a set of permutations g is the smallest permutation group containing g or equivalently, the set of products*

$$G = \bigcup_{i=0}^{\infty} g^i .$$

For computational purposes this description is very inefficient. A simple algorithm that allows one to compute the elements of the permutation group generated by g, provided that the degree of g is small, is given in the exercises.

Example 3.3
Let us look at more permutation groups of degree 4. First there are the cyclic permutation groups. In S_4 there are five different types of permutations: the identity, 2-cycles, 3-cycles, 4-cycles, and products of two disjoint 2-cycles. The last have order 2. Noncyclic groups that have already been mentioned are S_4 itself and A_4. There are also four copies of S_3: take the full group of permutations of any three of 1, 2, 3, 4. As well there is the permutation group V' mentioned at the beginning of the chapter. Another permutation group of order 4, which looks similar to V', is

$$V = \{(1), (1\,2)(3\,4), (1\,3)(2\,4), (1\,4)(2\,3)\} .$$

All nontrivial elements have order 2 and any two of them generate V. Notice that $V \subset A_4$. There is also a permutation group of order 8 called D_4, generated by $\{(1\,2\,3\,4), (1\,2)(3\,4)\}$. This is in fact the permutation group corresponding to the symmetries of a square, mentioned at the beginning of the chapter. There are actually several copies of this permutation group, depending on the choice of 4-cycle you make.   ▯

Software and Calculations

The package "Groups.m" can be used to make useful calculations in permutation groups. To start, you must load it:

```
In[1] := « Groups.m;
```

Next you have to know how our notation for permutations is implemented. In this package, a permutation (in mapping notation) is given by the list of its images with the header, M. For example

$$\begin{pmatrix} 1\,2\,3\,4\,5\,6 \\ 2\,4\,1\,6\,3\,5 \end{pmatrix}$$

is represented by

```
M[2,4,1,6,3,5]
```

You can name this permutation:

```
In[2] := a = M[2,4,1,6,3,5]
```

$$Out[2] = \begin{pmatrix} 1 & 2 & 3 & 4 & 5 & 6 \\ 2 & 4 & 1 & 6 & 3 & 5 \end{pmatrix}$$

If b is another permutation, then the product ab is given by

$$a.b$$

For example if

```
In[3] := b = M[4,2,3,6,1,5]
```

$$Out[3] = \begin{pmatrix} 1 & 2 & 3 & 4 & 5 & 6 \\ 4 & 2 & 3 & 6 & 1 & 5 \end{pmatrix}$$

then entering

```
In[4] := a.b
```

$$Out[4] = \begin{pmatrix} 1 & 2 & 3 & 4 & 5 & 6 \\ 6 & 4 & 1 & 5 & 2 & 3 \end{pmatrix}$$

To find the inverse of a, enter

```
In[5] := Inverse[a]
```

$$Out[5] = \begin{pmatrix} 1 & 2 & 3 & 4 & 5 & 6 \\ 3 & 1 & 5 & 2 & 6 & 4 \end{pmatrix}$$

A permutation in cycle notation is written as the list of its cycles, which are in turn lists, and is preceded by the header P. For example (1 2)(3 4 5) is represented by

$$P[\{1,2\},\{3,4,5\}]$$

You can take products and inverses using cycle notation just as with mapping notation:

```
In[6] := P[{1,2},{3,4,5}].P[{1,2,3,4,5}]
```

```
Out[6] = (2  4  3  5)
```

```
In[7]:= Inverse[ P[{1,2},{3,4,5}] ]
Out[7]= (1  2)(3  5  4)
```

Output is in standard mathematical cycle notation.

You can apply these functions to lists of permutations. When you take the product of two lists of permutations, the product of every permutation in the first list with every one in the second will be computed: for example

```
In[8]:= {a,b}.{a,b}
```

$$Out[8]= \left\{ \begin{pmatrix} 1 & 2 & 3 & 4 & 5 & 6 \\ 2 & 6 & 4 & 5 & 3 & 1 \end{pmatrix}, \begin{pmatrix} 1 & 2 & 3 & 4 & 5 & 6 \\ 4 & 6 & 2 & 5 & 1 & 3 \end{pmatrix}, \right.$$
$$\left. \begin{pmatrix} 1 & 2 & 3 & 4 & 5 & 6 \\ 6 & 2 & 3 & 5 & 4 & 1 \end{pmatrix}, \begin{pmatrix} 1 & 2 & 3 & 4 & 5 & 6 \\ 6 & 4 & 1 & 5 & 2 & 3 \end{pmatrix} \right\}$$

These functions are useful for checking whether a set of permutations form a group. For example, let

```
In[9]:= G = {M[2,3,1,5,4], M[4,1,2,5,3],
             M[1,3,5,2,4], M[2,1,3,4,5],
             M[1,2,3,4,5]}
```

$$Out[9]= \left\{ \begin{pmatrix} 1 & 2 & 3 & 4 & 5 \\ 2 & 3 & 1 & 5 & 4 \end{pmatrix}, \begin{pmatrix} 1 & 2 & 3 & 4 & 5 \\ 4 & 1 & 2 & 5 & 3 \end{pmatrix}, \begin{pmatrix} 1 & 2 & 3 & 4 & 5 \\ 1 & 3 & 5 & 2 & 4 \end{pmatrix}, \right.$$
$$\left. \begin{pmatrix} 1 & 2 & 3 & 4 & 5 \\ 2 & 1 & 3 & 4 & 5 \end{pmatrix}, \begin{pmatrix} 1 & 2 & 3 & 4 & 5 \\ 1 & 2 & 3 & 4 & 5 \end{pmatrix} \right\}$$

```
In[10]:= Inverse[G]
```

$$Out[10]= \left\{ \begin{pmatrix} 1 & 2 & 3 & 4 & 5 \\ 3 & 1 & 2 & 5 & 4 \end{pmatrix}, \begin{pmatrix} 1 & 2 & 3 & 4 & 5 \\ 2 & 3 & 5 & 1 & 4 \end{pmatrix}, \begin{pmatrix} 1 & 2 & 3 & 4 & 5 \\ 1 & 4 & 2 & 5 & 3 \end{pmatrix}, \right.$$
$$\left. \begin{pmatrix} 1 & 2 & 3 & 4 & 5 \\ 2 & 1 & 3 & 4 & 5 \end{pmatrix}, \begin{pmatrix} 1 & 2 & 3 & 4 & 5 \\ 1 & 2 & 3 & 4 & 5 \end{pmatrix} \right\}$$

So

$$\begin{pmatrix} 1 & 2 & 3 & 4 & 5 \\ 2 & 3 & 1 & 5 & 4 \end{pmatrix}^{-1} = \begin{pmatrix} 1 & 2 & 3 & 4 & 5 \\ 3 & 1 & 2 & 5 & 4 \end{pmatrix}$$

is not in the set G and therefore G is not a permutation group.

The function Group calculates the permutation group generated by a given set of permutations. All computations take place in S_n where n is the largest number occurring in the cycles in the input. The algorithm used is the one given in Exercise 3.8. For example, the permutation group $G = \langle (1\,2\,3\,4\,5), (1\,2)(3\,5) \rangle$ is given by

```
In[11]:= G = Group[ P[{1, 2, 3, 4, 5}] ,
              P[{1,2},{3,5}] ]

Out[11]= ⟨(1 2 3 4 5),(1 2)(3 5)⟩
```

To see a list of the elements in G, you use the function Elements:

```
In[12]:= Elements[G]

Out[12]= {(1), (1 2 3 4 5), (1 3 5 2 4), (1 4 2 5 3),
          (1 5 4 3 2), (1 2)(3 5), (1 3)(4 5),
          (1 4)(2 3), (1 5)(2 4), (2 5)(3 4)}
```

The function Generators will print out the generators of G again:

```
In[13]:= Generators[G]

Out[13]= {(1 2 3 4 5), (1 2)(3 5)}
```

The order of the permutation group is given by the function Order:

```
In[14]:= Order[G]

Out[14]= 10
```

We can verify by computation that (1 2) and (1 2 3 4 5 6) generate S_6:

```
In[15]:= Order[ Group[ P[1,2], P[1,2,3,4,5,6] ] ]

Out[15]= 720
```

Since $6! = 720$, they do generate S_6.

Let us look for generators of A_4, which is also not cyclic. We know that the even permutations of degree 4 are the 3-cycles, the products of two disjoint transpositions and the identity. We first try two 3-cycles:

```
In[16]:= Order[ Group[ P[{1,2,3}], P[{2,3,4}] ] ]

Out[16]= 12
```

It is not hard to show that the order of A_n is $n!/2$ (see Exercise 3.11). Therefore the two 3-cycles do generate A_4. How about a 3-cycle and a product of two disjoint transpositions?

```
In[17]:= Order [ Group[ P[{1,2,3}], P[{1,2},{3,4}] ] ]
```

Out [17] = 12

These two appear to generate A_4 as well. Let us see how these calculations extend to A_5. We could use two 3-cycles again. But that would not work in A_6. So we shall try with a 5-cycle and a 3-cycle:

In [18] := Order [Group [P[{1,2,3,4,5}], P[{1,2,3}]]]

Out [18] = 60

And 60 is the order of A_5. We now try a 5-cycle and a product of two disjoint transpositions:

In [19] := Order[Group [P[{1,2,3,4,5}],
                    P[{1,2},{3,4}]]]

Out [19] = 60

This works too. Do these calculations extend to A_6? Do they generalize to A_n, for arbitrary n?

Exercises

3.1 Does the set of all odd permutations form a permutation group?

3.2 Verify the identity

$$\alpha^{-i} = \alpha^{r-i} \quad \text{for any } i, 1 \leq i < r .$$

for a permutation α of order r.

3.3 Verify the identity

$$(i\ j) = (j-1\ j)(j-2\ j-1) \cdots (i\ i+1) \cdots (j-2\ j-1)(j-1\ j) .$$

for $i < j$.

3.4 Show that the set of $n - 1$ transpositions

$$\{(1\ 2), (1\ 3), \cdots , (1\ n)\}$$

generates S_n.

3.5 We have mentioned permutation groups of degree 3, of order 1, 2, 3, and 6. Do any exist of order 4 or 5?

3.6 Show that the permutation group generated by a set of permutations $g \subset S_n$ is the smallest set $G \subset S_n$ containing g such that $gG \subset G$.

3.7 Apply the algorithm in the exercise below to the set

$$g = \{(1\,2\,3), (2\,3\,4)\} \,.$$

List the values of h and G at each stage.

3.8 Prove that the following algorithm terminates with G, the permutation group generated by g (cf. [3], Exercise 3.3.13).

(a) Set $G = \{(1)\}$ and $h = \{(1)\}$.

(b) Set $h = hg \setminus G$.

(c) If $h = \emptyset$, stop.

(d) Set $G = G \cup h$ and go to step (b).

3.9 Check whether the following sets of permutations form permutation groups.

(a) A={M[2,3,1,5,4],M[4,1,2,5,3],M[1,3,5,2,4],M[2,1,3,4,5], M[1,2,3,4,5]}

(b) B={M[1,2,3,4],M[2,1,4,3],M[3,4,1,2],M[4,3,2,1]}

(c) G={M[1, 2, 3, 4, 5], M[1, 5, 4, 3, 2], M[2, 1, 5, 4, 3], M[2, 3, 4, 5, 1], M[3, 2, 1, 5, 4], M[3, 4, 5, 1, 2], M[4, 3, 2, 1, 5], M[4, 5, 1, 2, 3], M[5, 1, 2, 3, 4], M[5, 4, 3, 2, 1]}

3.10 (a) Find as many different types of permutation groups of degree 5 as you can. Describe them in terms of generators rather than listing all the elements.

(b) Make a list of the orders of all the permutation groups you found in (a). What do these integers have in common?

3.11 What is the order of A_n?

3.12 • Verify that the set of 3-cycles generates A_4, A_5, and A_6. Can you prove that this is true for any $n \geq 3$?

3.13 (a) Find two permutations that generate A_6. Find two that generate A_7.

(b) Do your results generalize to A_n for any n? Make a conjecture and try to prove it.

Chapter 4

Linear Groups

Definitions and Examples

Think of the set of all rotations about the origin in the Euclidean plane. Let $\alpha(t)$ denote the rotation through the angle t counterclockwise. It can be represented by the matrix

$$\begin{pmatrix} \cos t & -\sin t \\ \sin t & \cos t \end{pmatrix}.$$

If we multiply two such rotations, we get another rotation, and the inverse of a rotation is also a rotation. In fact,

$$\alpha(t)\alpha(t') = \alpha(t + t') \qquad \alpha(t)^{-1} = \alpha(-t).$$

So if we set

$$G = \{\alpha(t) \mid t \in \mathbb{R}\},$$

we get a collection of real 2×2 matrices that has the same algebraic properties as a permutation group. Such sets of square matrices are called *linear groups*. We want to allow matrices with coefficients in an arbitrary field.

An $m \times n$ matrix defined over a field F is a rectangular array

$$\begin{pmatrix} a_{11} & a_{12} & \cdots & a_{1n} \\ a_{21} & a_{22} & \cdots & a_{2n} \\ \vdots & \vdots & \ddots & \vdots \\ a_{m1} & a_{m2} & \cdots & a_{mn} \end{pmatrix}$$

where $a_{ij} \in F$ for $1 \le i \le m$, $1 \le j \le n$. Matrix multiplication and addition can be defined just as in the case $F = \mathbb{R}$. And they have the same formal properties. We denote by $M(m, n, F)$ the set of all $m \times n$ matrices over F, by $M(n, F)$ the set of all $n \times n$ matrices, and by $GL(n, F)$ the subset of all invertible matrices in $M(n, F)$.

DEFINITION 4.1 *A linear group G is a nonempty set of matrices in $GL(n, F)$ such that*

(i) if $\alpha, \beta \in G$, then $\alpha\beta \in G$;

(ii) if $\alpha \in G$, then $\alpha^{-1} \in G$.

Thus $GL(n, F)$ itself is a linear group, called the *general linear group* (of degree n over F). Another simple example is the *special linear group*, which is defined by

$$SL(n, F) := \{\alpha \in GL(n, F) \mid \det \alpha = 1\}.$$

The determinant of a matrix $\alpha \in M(n, F)$ is defined just as for $F = \mathbb{R}$ and has the same basic properties. Now for $\alpha, \beta \in SL(n, F)$,

$$\det(\alpha\beta) = \det \alpha \ \det \beta = 1$$
$$\det(\alpha^{-1}) = (\det \alpha)^{-1} = 1.$$

Thus $SL(n, F)$ is also a linear group.

We are particularly interested in linear groups over the finite fields $\mathbb{F}_p$. If a linear group G is finite then the *order* of G, written $|G|$, is defined to be the number of elements of G. If G is infinite, we write $|G| = \infty$.

Examples 4.1

(i) For any field F let

$$T = \left\{ \begin{pmatrix} 1 & b \\ 0 & 1 \end{pmatrix} \middle| b \in F \right\} \subset GL(2, F).$$

There is a one-to-one correspondence between F itself and T. Under this correspondence, addition in F corresponds to matrix multiplication in T. It follows that T is a linear group.

(ii) Let $N(p)$ be the set of upper triangular matrices in $GL(2, \mathbb{F}_p)$,

$$N(p) = \left\{ \begin{pmatrix} a & b \\ 0 & d \end{pmatrix} \middle| a, b, d \in \mathbb{F}_p, ad \neq 0 \right\} \subset GL(2, \mathbb{F}_p).$$

It is easy to see that $N(p)$ is a linear group of order $p(p-1)^2$.

(iii) Let

$$G(p) = \left\{ \begin{pmatrix} a & b \\ br & a \end{pmatrix} \middle| a, b \in \mathbb{F}_p, a^2 - b^2 r \neq 0 \right\} \subset GL(2, \mathbb{F}_p)$$

where $r \in \mathbb{F}_p$ is not a square. It is easy to check that the product of two matrices in $G(p)$ lies in $G(p)$, as does the inverse of any matrix in $G(p)$. So $G(p)$ is a linear group.

To compute its order we have to know precisely which matrices have nonzero determinants. Now the determinant of such a matrix is $a^2 - b^2 r$. First we solve the congruence

$$a^2 - b^2 r \equiv 0 \pmod{p} .$$

If $b \not\equiv 0 \pmod{p}$, then multiplying by b^{-2} we get

$$(ab^{-1})^2 - r \equiv 0 \pmod{p} .$$

But then $x = ab^{-1}$ would be a solution of

$$x^2 \equiv r \pmod{p} .$$

This has no solution since r is not a square mod p. Thus we must have $b \equiv 0$ (mod p), which means that $a^2 \equiv 0 \pmod{p}$ and hence $a \equiv 0 \pmod{p}$. Therefore the only solution of our original quadratic congruence is the trivial one, $a = b = 0$. It follows that

$$G(p) = \left\{ \begin{pmatrix} a & b \\ br & a \end{pmatrix} \middle| a, b \in \mathbb{F}_p, \ a \neq 0 \text{ or } b \neq 0 \right\} .$$

This tells us that

$$|G| = p^2 - 1 .$$

(iv) Let us compute the order of $GL(2, \mathbb{F}_p)$. The number of matrices in $M(2, \mathbb{F}_p)$ is p^4. One of these is in $GL(2, \mathbb{F}_p)$ if and only if it has rank 2. We need to determine how many matrices have rank less than 2. A matrix has rank less than 2 if and only if its column vectors are linearly dependent, i.e., if and only if one is a multiple of the other. Now for the first column there are p^2 possible choices. For each of these, except $(0, 0)$, there are p distinct multiples, i.e., possibilities for the second column. This gives $(p^2 - 1)p$ possible matrices in all. If the first column is $(0, 0)$, then the second column can be any vector. This gives another p^2 matrices. So in all there are

$$(p^2 - 1)p + p^2 = p^3 + p^2 - p$$

matrices of rank less than 2. Therefore the order of $GL(2, \mathbb{F}_p)$ is

$$p^4 - p^3 - p^2 + p = (p - 1)^2 p(p + 1) . \qquad \Box$$

Generators

As for permutations, we can define the order $|\alpha|$ of an element α of a linear group to be the smallest $n \in \mathbb{N}$ such that $\alpha^n = 1$. If there is no such n, then we say that α has infinite order, and write $|\alpha| = \infty$. In example (i), let

$$\alpha = \begin{pmatrix} 1 & 1 \\ 0 & 1 \end{pmatrix} .$$

Then

$$\alpha^b = \begin{pmatrix} 1 & b \\ 0 & 1 \end{pmatrix}$$

for any $b \in \mathbb{Z}$. If $F = \mathbb{F}_p$, then

$$|\alpha| = p \ .$$

But if $F = \mathbb{Q}$, then

$$|\alpha| = \infty \ .$$

Two more examples: for any field F,

$$\begin{pmatrix} 0 & 1 \\ 1 & 0 \end{pmatrix} \in GL(2, F)$$

has order 2. In $GL(2, \mathbb{F}_5)$,

$$\begin{pmatrix} 1 & 0 \\ 0 & 2 \end{pmatrix}$$

has order 4 because $2^4 = 1$ in $\mathbb{F}_5$, but $2^2, 2^3 \neq 1$.

If the order of a matrix α is infinite, then α^{-1} is not a positive power of α. So in defining a cyclic linear group, we must include powers of α^{-1}.

DEFINITION 4.2   *Let $\alpha \in GL(n, F)$ for some field F. The cyclic linear group generated by α is the linear group $\langle \alpha \rangle := \{\alpha^b \mid b \in \mathbb{Z}\}$.*

As in permutation groups, we see that the order of an element α is the order of $\langle \alpha \rangle$. In general we say that a linear group G is generated by a set $g \subset G$ if every element of G can be written as a product of elements of g and their inverses.

Example 4.2

Using row and column operations we can find generators for $GL(2, F)$. These operations work over an arbitrary field F in exactly the same way as over $\mathbb{R}$. We begin with a matrix

$$\begin{pmatrix} a & b \\ c & d \end{pmatrix}, \quad a, b, c, d \in F, \ ad - bc \neq 0 \ .$$

By interchanging the rows or the columns, we can ensure that $a \neq 0$. This means we multiply on the left or on the right by the matrix

$$\begin{pmatrix} 0 & 1 \\ 1 & 0 \end{pmatrix} \ .$$

Next, we add a multiple of the first row to the second to eliminate the entry c. This means multiplying on the left by a matrix of the form

$$\begin{pmatrix} 1 & 0 \\ x & 1 \end{pmatrix}, \quad x \in F \ .$$

We now have a matrix that looks like

$$\begin{pmatrix} a & b \\ 0 & d \end{pmatrix} \ , \ ad \neq 0 \ .$$

The next step is to eliminate the entry b. To do this, we add a multiple of the first column to the second. This means multiplying on the right by a matrix

$$\begin{pmatrix} 1 & y \\ 0 & 1 \end{pmatrix} \ , \ y \in F \ .$$

We are left with a diagonal matrix that looks like

$$\begin{pmatrix} a & 0 \\ 0 & d \end{pmatrix} = \begin{pmatrix} a & 0 \\ 0 & 1 \end{pmatrix} \begin{pmatrix} 1 & 0 \\ 0 & d \end{pmatrix} \ , \ a, d \in F \setminus \{0\} \ .$$

So any matrix in $GL(2, F)$ can be written as a product of matrices

$$\begin{pmatrix} a & 0 \\ 0 & 1 \end{pmatrix} , \ \begin{pmatrix} 1 & 0 \\ 0 & d \end{pmatrix} , \ \begin{pmatrix} 1 & 0 \\ c & 1 \end{pmatrix} , \ \begin{pmatrix} 1 & b \\ 0 & 1 \end{pmatrix} , \ \begin{pmatrix} 0 & 1 \\ 1 & 0 \end{pmatrix} \ , \ a, d \in F \setminus \{0\} \ , \ b, c \in F \ .$$

But we can do better:

$$\begin{pmatrix} 0 & 1 \\ 1 & 0 \end{pmatrix} \begin{pmatrix} 1 & c \\ 0 & 1 \end{pmatrix} \begin{pmatrix} 0 & 1 \\ 1 & 0 \end{pmatrix} = \begin{pmatrix} 1 & 0 \\ c & 1 \end{pmatrix}$$

and

$$\begin{pmatrix} 0 & 1 \\ 1 & 0 \end{pmatrix} \begin{pmatrix} d & 0 \\ 0 & 1 \end{pmatrix} \begin{pmatrix} 0 & 1 \\ 1 & 0 \end{pmatrix} = \begin{pmatrix} 1 & 0 \\ 0 & d \end{pmatrix} \ .$$

So all we need are matrices

$$\begin{pmatrix} a & 0 \\ 0 & 1 \end{pmatrix} , \ \begin{pmatrix} 1 & b \\ 0 & 1 \end{pmatrix} , \ \begin{pmatrix} 0 & 1 \\ 1 & 0 \end{pmatrix} \ , \ a \in F \setminus \{0\} \ , \ b \in F \ .$$

This is our set of generators of $GL(2, F)$, for any field F. If $F = \mathbb{F}_p$, then we can reduce the number of generators further. As we mentioned before

$$\begin{pmatrix} 1 & b \\ 0 & 1 \end{pmatrix} = \begin{pmatrix} 1 & 1 \\ 0 & 1 \end{pmatrix}^b$$

for $b \in \mathbb{Z}$. Thus

$$\left\{ \begin{pmatrix} 1 & 1 \\ 0 & 1 \end{pmatrix} , \begin{pmatrix} 0 & 1 \\ 1 & 0 \end{pmatrix} , \begin{pmatrix} a & 0 \\ 0 & 1 \end{pmatrix} \ \middle| \ a \in \mathbb{F}_p \setminus \{0\} \right\}$$

generates $GL(2, \mathbb{F}_p)$. It can be shown that

$$\left\{ \begin{pmatrix} a & 0 \\ 0 & 1 \end{pmatrix} \ \middle| \ a \in F \setminus \{0\} \right\}$$

is a cyclic linear group (see Chapter 14). So in fact you can find a set of three matrices that generates $GL(2, \mathbb{F}_p)$. For example, take $p = 5$. In $\mathbb{F}_5$ we have $2^2 = 4$,

$2^3 = 3, 2^4 = 1$. Thus every nonzero element of $\mathbb{F}_5$ is a power of 2. So every matrix of the form

$$\begin{pmatrix} a & 0 \\ 0 & 1 \end{pmatrix} = \begin{pmatrix} 2 & 0 \\ 0 & 1 \end{pmatrix}^i$$

for some i. Therefore

$$\left\{ \begin{pmatrix} 1 & 1 \\ 0 & 1 \end{pmatrix}, \begin{pmatrix} 0 & 1 \\ 1 & 0 \end{pmatrix}, \begin{pmatrix} 2 & 0 \\ 0 & 1 \end{pmatrix} \right\} \tag{4.1}$$

generates $GL(2, \mathbb{F}_5)$. □

This discussion has actually given us an algorithm for expressing a matrix in $GL(2, F)$ in terms of these generators. Let us apply it to the matrix

$$\begin{pmatrix} 0 & 2 \\ 3 & 1 \end{pmatrix} \in GL(2, \mathbb{F}_5) .$$

To begin we must have a nonzero entry in the upper left-hand corner. So we will interchange columns, or equivalently, multiply the given matrix on the right by

$$\begin{pmatrix} 0 & 1 \\ 1 & 0 \end{pmatrix} .$$

This gives us

$$\begin{pmatrix} 0 & 2 \\ 3 & 1 \end{pmatrix} \begin{pmatrix} 0 & 1 \\ 1 & 0 \end{pmatrix} = \begin{pmatrix} 2 & 0 \\ 1 & 3 \end{pmatrix} .$$

Next we add 2 (row 1) to row 2 to eliminate the first entry:

$$\begin{pmatrix} 1 & 0 \\ 2 & 1 \end{pmatrix} \begin{pmatrix} 2 & 0 \\ 1 & 3 \end{pmatrix} = \begin{pmatrix} 2 & 0 \\ 0 & 3 \end{pmatrix} .$$

So

$$\begin{pmatrix} 1 & 0 \\ 2 & 1 \end{pmatrix} \begin{pmatrix} 0 & 2 \\ 3 & 1 \end{pmatrix} \begin{pmatrix} 0 & 1 \\ 1 & 0 \end{pmatrix} = \begin{pmatrix} 2 & 0 \\ 0 & 3 \end{pmatrix} ,$$

or

$$\begin{pmatrix} 0 & 2 \\ 3 & 1 \end{pmatrix} = \begin{pmatrix} 1 & 0 \\ 2 & 1 \end{pmatrix}^{-1} \begin{pmatrix} 2 & 0 \\ 0 & 3 \end{pmatrix} \begin{pmatrix} 0 & 1 \\ 1 & 0 \end{pmatrix}^{-1}$$

$$= \begin{pmatrix} 1 & 0 \\ 3 & 1 \end{pmatrix} \begin{pmatrix} 2 & 0 \\ 0 & 3 \end{pmatrix} \begin{pmatrix} 0 & 1 \\ 1 & 0 \end{pmatrix} . \tag{4.2}$$

Now

$$\begin{pmatrix} 1 & 0 \\ 3 & 1 \end{pmatrix} = \begin{pmatrix} 0 & 1 \\ 1 & 0 \end{pmatrix} \begin{pmatrix} 1 & 3 \\ 0 & 1 \end{pmatrix} \begin{pmatrix} 0 & 1 \\ 1 & 0 \end{pmatrix}$$

$$= \begin{pmatrix} 0 & 1 \\ 1 & 0 \end{pmatrix} \begin{pmatrix} 1 & 1 \\ 0 & 1 \end{pmatrix}^3 \begin{pmatrix} 0 & 1 \\ 1 & 0 \end{pmatrix} .$$

And

$$
\begin{pmatrix} 2 & 0 \\ 0 & 3 \end{pmatrix} = \begin{pmatrix} 2 & 0 \\ 0 & 1 \end{pmatrix} \begin{pmatrix} 1 & 0 \\ 0 & 3 \end{pmatrix}
$$

$$
= \begin{pmatrix} 2 & 0 \\ 0 & 1 \end{pmatrix} \begin{pmatrix} 0 & 1 \\ 1 & 0 \end{pmatrix} \begin{pmatrix} 3 & 0 \\ 0 & 1 \end{pmatrix} \begin{pmatrix} 0 & 1 \\ 1 & 0 \end{pmatrix}
$$

$$
= \begin{pmatrix} 2 & 0 \\ 0 & 1 \end{pmatrix} \begin{pmatrix} 0 & 1 \\ 1 & 0 \end{pmatrix} \begin{pmatrix} 2 & 0 \\ 0 & 1 \end{pmatrix}^3 \begin{pmatrix} 0 & 1 \\ 1 & 0 \end{pmatrix} .
$$

Finally, we can substitute the last two calculations into Equation (4.2):

$$
\begin{pmatrix} 0 & 2 \\ 3 & 1 \end{pmatrix} = \begin{pmatrix} 0 & 1 \\ 1 & 0 \end{pmatrix} \begin{pmatrix} 1 & 1 \\ 0 & 1 \end{pmatrix}^3 \begin{pmatrix} 0 & 1 \\ 1 & 0 \end{pmatrix} \begin{pmatrix} 2 & 0 \\ 0 & 1 \end{pmatrix} \begin{pmatrix} 0 & 1 \\ 1 & 0 \end{pmatrix} \begin{pmatrix} 2 & 0 \\ 0 & 1 \end{pmatrix}^3 .
$$

So we have now written the given matrix in terms of the generators (4.1).

Given $g \subset GL(n, F)$, for some field F and some natural number n, the linear group generated by g, written $\langle g \rangle$, is the smallest linear group containing g. If F is finite then Definition 3.3 and the remarks with it apply, as does the algorithm for computing $\langle g \rangle$.

Software and Calculations

The package "Groups.m" also allows you to make calculations with linear groups over a field $\mathbb{F}_p$. After loading the package, you must first choose a prime with the function ChoosePrime . For example

```
In[1] := « Groups.m;

In[2] := ChoosePrime[11]

Out[2] = 11
```

All your calculations will then be modulo this prime. You can change the prime at any time, but if you do not make a choice none of the functions will evaluate. Matrices are usually represented in *Mathematica* as lists of lists. In this package, however, such a representation is wrapped with L [] in order to give a mechanism for reducing modulo the chosen prime at each step in the calculations. (As well it avoids confusion with permutations in cycle notation that are also represented in *Mathematica* as lists of lists.) For example, the general 2×2 matrix

$$
\begin{pmatrix} a & b \\ c & d \end{pmatrix} ,
$$

which is usually represented in *Mathematica* as

$$\{\{a, b\}, \{c, d\}\},$$

is represented in this package as

$$L[\{a, b\}, \{c, d\}].$$

The same functions you used with permutation groups are defined for linear groups. For example if

In[3]: = a = L[{1,2},{3,4}]

Out[3] = $\begin{pmatrix} 1 & 2 \\ 3 & 4 \end{pmatrix}$

In[4]: = b = L[{6,3},{7,5}]

Out[4] = $\begin{pmatrix} 6 & 3 \\ 7 & 5 \end{pmatrix}$

then

In[5]: = Inverse[a]

Out[5] = $\begin{pmatrix} 9 & 1 \\ 7 & 5 \end{pmatrix}$

In[6]: = a.b

Out[6] = $\begin{pmatrix} 9 & 2 \\ 2 & 7 \end{pmatrix}$

Now suppose

In[7]: = ChoosePrime[7]

Out[7] = 7

and

$In[8] := G = \{ L[\{1,1\},\{1,2\}], L[\{2,3\},\{4,1\}],$
$\qquad\qquad L[\{2,6\},\{6,1\}], L[\{1,0\},\{0,1\}],$
$\qquad\qquad L[\{2,1\},\{6,4\}] \}$

$Out[8] = \left\{ \begin{pmatrix} 1 & 1 \\ 1 & 2 \end{pmatrix}, \begin{pmatrix} 2 & 3 \\ 4 & 1 \end{pmatrix}, \begin{pmatrix} 2 & 6 \\ 6 & 1 \end{pmatrix}, \begin{pmatrix} 1 & 0 \\ 0 & 1 \end{pmatrix}, \begin{pmatrix} 2 & 1 \\ 6 & 4 \end{pmatrix} \right\}$

Then

$In[9] := Inverse[G]$

$Out[9] = \left\{ \begin{pmatrix} 2 & 6 \\ 6 & 1 \end{pmatrix}, \begin{pmatrix} 2 & 1 \\ 6 & 4 \end{pmatrix}, \begin{pmatrix} 1 & 1 \\ 1 & 2 \end{pmatrix}, \begin{pmatrix} 1 & 0 \\ 0 & 1 \end{pmatrix}, \begin{pmatrix} 2 & 3 \\ 4 & 1 \end{pmatrix} \right\}$

So G is closed under taking inverses. Let us check whether it is closed under products:

$In[10] := G.G$

$Out[10] = \left\{ \begin{pmatrix} 0 & 5 \\ 2 & 5 \end{pmatrix}, \begin{pmatrix} 1 & 0 \\ 0 & 1 \end{pmatrix}, \begin{pmatrix} 1 & 1 \\ 0 & 4 \end{pmatrix}, \begin{pmatrix} 1 & 1 \\ 1 & 2 \end{pmatrix}, \begin{pmatrix} 1 & 5 \\ 0 & 2 \end{pmatrix}, \begin{pmatrix} 2 & 1 \\ 6 & 4 \end{pmatrix}, \right.$
$\qquad\qquad \begin{pmatrix} 2 & 2 \\ 5 & 6 \end{pmatrix}, \begin{pmatrix} 2 & 3 \\ 3 & 5 \end{pmatrix}, \begin{pmatrix} 2 & 3 \\ 4 & 1 \end{pmatrix}, \begin{pmatrix} 2 & 6 \\ 6 & 1 \end{pmatrix}, \begin{pmatrix} 3 & 4 \\ 3 & 0 \end{pmatrix}, \begin{pmatrix} 3 & 6 \\ 1 & 1 \end{pmatrix},$
$\qquad\qquad \left. \begin{pmatrix} 3 & 6 \\ 1 & 5 \end{pmatrix}, \begin{pmatrix} 5 & 1 \\ 5 & 6 \end{pmatrix}, \begin{pmatrix} 5 & 4 \\ 4 & 2 \end{pmatrix}, \begin{pmatrix} 5 & 5 \\ 4 & 3 \end{pmatrix}, \begin{pmatrix} 6 & 4 \\ 3 & 5 \end{pmatrix} \right\}$

You see immediately that the product of $L[\{1,1\},\{1,2\}]$ with itself is not in G. So G is not a linear group.

We can verify that $GL(2, \mathbb{F}_5)$ is generated by the three matrices given in the previous section. First we change the chosen prime to 5.

$In[11] := ChoosePrime[5]$

$Out[11] = 5$

Let

$In[12] := H = Group[L[\{\{2,0\},\{0,1\}\}], L[\{\{0,1\},\{1,0\}\}],$
$\qquad\qquad L[\{\{1,1\},\{0,1\}\}]]$

$$\text{Out [12]} = \left\langle \begin{pmatrix} 2 & 0 \\ 0 & 1 \end{pmatrix}, \begin{pmatrix} 0 & 1 \\ 1 & 0 \end{pmatrix}, \begin{pmatrix} 1 & 1 \\ 0 & 1 \end{pmatrix} \right\rangle$$

Now we check:

In [13] := Order [H]

Out [13] = 480

This is indeed $4^2 \cdot 5 \cdot 6$. Can you find three matrices that generate $GL(2, \mathbb{F}_7)$? $GL(2, \mathbb{F}_{11})$?

Exercises

4.1 Verify that $G(p)$ is a linear group.

4.2 Show that for any field F the set N of invertible 2×2 upper triangular matrices forms a linear group. Verify that if $F = \mathbb{F}_p$ then $|N(p)| = p(p-1)^2$.

4.3 Show that in T

$$\begin{pmatrix} 1 & b \\ 0 & 1 \end{pmatrix}\begin{pmatrix} 1 & b' \\ 0 & 1 \end{pmatrix} = \begin{pmatrix} 1 & b+b' \\ 0 & 1 \end{pmatrix} \quad \text{and} \quad \begin{pmatrix} 1 & b \\ 0 & 1 \end{pmatrix}^{-1} = \begin{pmatrix} 1 & -b \\ 0 & 1 \end{pmatrix}$$

for any $b, b' \in F$.

4.4 • Let

$$F_{p(p-1)} = \left\{ \begin{pmatrix} a & b \\ 0 & 1 \end{pmatrix} \middle| a, b \in \mathbb{F}_p, \ a \neq 0 \right\} \subset GL(2, \mathbb{F}_p).$$

Check that $F_{(p-1)p}$ is a linear group (called the Frobenius group). What is its order? Verify that

$$\left\{ \begin{pmatrix} 1 & 1 \\ 0 & 1 \end{pmatrix}, \begin{pmatrix} 2 & 0 \\ 0 & 1 \end{pmatrix} \right\} \subset GL(2, \mathbb{F}_5)$$

generates F_{20}.

4.5 • In $SL(2, \mathbb{C})$ let

$$Q = \left\{ \pm \begin{pmatrix} 1 & 0 \\ 0 & 1 \end{pmatrix}, \pm \begin{pmatrix} i & 0 \\ 0 & -i \end{pmatrix}, \pm \begin{pmatrix} 0 & 1 \\ -1 & 0 \end{pmatrix}, \pm \begin{pmatrix} 0 & i \\ i & 0 \end{pmatrix} \right\}.$$

Verify that Q is a linear group. What are the orders of its 8 elements? Q is called the quaternion group.

4.6 Write the matrix

$$\begin{pmatrix} 1 & 2 \\ 3 & 4 \end{pmatrix} \in GL(2, \mathbb{F}_5)$$

in terms of the generators above.

4.7 Find three matrices that generate $GL(2, \mathbb{F}_7)$. Check your answer with the software functions.

4.8 • Prove that for any field F, $SL(2, F)$ is generated by

$$\left\{ \begin{pmatrix} a & 0 \\ 0 & a^{-1} \end{pmatrix}, \begin{pmatrix} 1 & b \\ 0 & 1 \end{pmatrix}, \begin{pmatrix} 0 & 1 \\ -1 & 0 \end{pmatrix} \middle| a \in F \setminus \{0\}, b \in F \right\}.$$

4.9 Find three matrices that generate $SL(2, \mathbb{F}_{11})$.

4.10 Compute $|SL(2, \mathbb{F}_7)|$ and $|SL(2, \mathbb{F}_{11})|$. Can you give a formula for $|SL(2, \mathbb{F}_p)|$ in general?

4.11 • Find an element of order 3 in $SL(2, \mathbb{F}_5)$. Find a linear group $G \subset SL(2, \mathbb{F}_5)$ of order 24.

Chapter 5

Groups

In Chapters 3 and 4 we discussed permutation groups and linear groups in a way that brought out the formal similarities between them. These similarities lead to the definition of an abstract group. First we must make clear what "multiplication" means in a general context and then what its basic properties should be.

Basic Properties and More Examples

DEFINITION 5.1   *A binary operation on a set G is a mapping: $G \times G \to G$.*

We write such an operation as a "multiplication," keeping in mind that it can mean such different operations as matrix multiplication or composition of permutations or addition of real numbers. The basic properties that our operation should have are those that it has in our two examples.

DEFINITION 5.2   *A set G with a binary operation on it is called a group if it has the following properties:*

(i) the operation is associative, i.e., for any $\alpha, \beta, \gamma \in G$

$$(\alpha\beta)\gamma = \alpha(\beta\gamma) ;$$

(ii) there is an identity element, written $1 \in G$, i.e., for any $\alpha \in G$

$$\alpha \cdot 1 = 1 \cdot \alpha = \alpha ;$$

(iii) every element in G has an inverse, i.e., for every $\alpha \in G$ there exists an element written $\alpha^{-1} \in G$ such that

$$\alpha \cdot \alpha^{-1} = \alpha^{-1} \cdot \alpha = 1 .$$

The binary operation on G is called the group operation.

Examples 5.1

(i) Take G to be a permutation group with composition as the group operation, as explained in Chapter 3. By the definition of a permutation group, this is an operation in the above sense. It is certainly associative. The identity element is the identity permutation (1). The inverse of a permutation is the inverse permutation, which belongs to G, again by definition.

(ii) Take G to be a linear group, with matrix multiplication as the group operation. The definition of a linear group ensures that this is a binary operation. It is associative. The identity element is the identity matrix I. Every matrix in G has an inverse in G by definition.

(iii) Let $G = \mathbb{Z}$ (or $\mathbb{Q}$ or $\mathbb{R}$ or $\mathbb{C}$) with the operation addition. Addition is associative, the identity element is 0, and the inverse of any $a \in \mathbb{Z}$ is $-a$. In fact, any field F is a group under addition, as is $\mathbb{Z}/n\mathbb{Z}$ for any $n \in \mathbb{N}$.

(iv) Let
$$F^{\times} := \{a \in F \mid a \neq 0\}$$

with the operation multiplication. It is associative, with identity element $1 \in F$, and the inverse of any $a \in F$ is a^{-1}.

(v) The nonzero integers do not form a group under multiplication because no integer except ± 1 has a multiplicative inverse.

(vi) For $m \in \mathbb{N}$, let
$$(\mathbb{Z}/m\mathbb{Z})^{\times} = \{\bar{a} \in \mathbb{Z}/m\mathbb{Z} \mid (a, m) = 1\} .$$

If $(a, m) = 1$ and $(b, m) = 1$, then $(ab, m) = 1$. Furthermore, if $(a, m) = 1$, then as we saw in Theorem 1.4, there exists an integer x such that $ax \equiv 1$ (mod m). For this congruence to hold, x must be relatively prime to m. Therefore $(\mathbb{Z}/m\mathbb{Z})^{\times}$ is a group under multiplication mod m, called the multiplicative group of $\mathbb{Z}/m\mathbb{Z}$. In the particular case where m is a prime p, we get $\mathbb{F}_p^{\times}$.

(vii) Let $G = \mathbb{R}^n$ with operation vector addition. It is associative, the vector $(0, 0, \ldots, 0)$ is the identity element, and the inverse of a vector $(a_1, a_2, \ldots, a_n)$ is $(-a_1, -a_2, \ldots, -a_n)$.

(viii) The space of matrices $M(m, n, F)$ under matrix addition is a group.

(ix) Let
$$S = \{a \in \mathbb{C} \mid |a| = 1\} .$$

Under multiplication S is a group.

(x) Let $GL(n, \mathbb{Z})$ be the set of all invertible $n \times n$ matrices with integer entries whose inverses also have integer entries. Under matrix multiplication $GL(n, \mathbb{Z})$ is a group, just like $GL(n, F)$ for a field F. If $\alpha \in GL(n, \mathbb{Z})$ then $\det(\alpha)$ and $\det(\alpha^{-1}) = \det(\alpha)^{-1}$ are integers. Since $\det(\alpha)\det(\alpha^{-1}) = 1$, they must both be ± 1.

(xi) A mapping $\alpha : \mathbb{R}^n \to \mathbb{R}^n$ is a *rigid motion* or *isometry* if it preserves distances, in other words,

$$\|\alpha(v) - \alpha(w)\| = \|v - w\|$$

for all $v, w \in \mathbb{R}^n$, where $\|v\|$ denotes the length of v (cf. [4]). The composition of two isometries is again one. It is easy to see that an isometry is injective. One can also show that it is surjective. Thus any isometry has an inverse, which is also an isometry. Let $Iso(n)$ denote the set of isometries. Composition is a binary operation on $Iso(n)$, which is associative. The identity mapping is an isometry. So $Iso(n)$ is a group.

(xii) The trivial group is the group with only one element, the identity element: $\{1\}$.

(xiii) Now for something completely different: the *braid group*. Imagine that we have two parallel lines. On the first one, on the right, we have n points labeled $1, 2, \ldots, n$. On the second, on the left, we also have n points $1, 2, \ldots, n$, which are just the translates of the corresponding points on the first. We join each point on the first line to one on the second line with a thread, going from right to left, in such a way that no two points on the first are joined to the same point on the second. This is called a *braid*.

We consider two such constructions to be the same braid if we can get one from the other by moving about the strands without changing their end points. So what matter are the end points and the way the strands cross each other. We can combine two braids by placing them end to end to get a new braid.

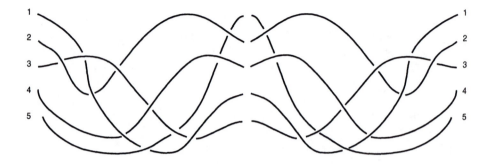

The trivial braid, in which 1 is joined to 1, 2 to 2, and so on, without any crossovers, is the identity element for this operation.

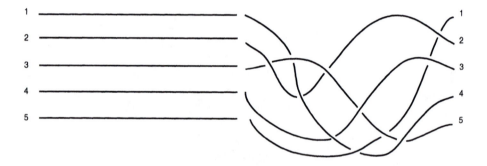

The inverse of a braid is just the braid obtained by running along the strands from their end points to their starting points. The braid below is the inverse of the braid we started with. The composition shown above is just the inverse composed with the original braid. You can see that by moving the strands, you will get the trivial braid.

With this operation, the set of braids, denoted by B_n, forms a group. It is a good idea to make a model with string, of braids with two strands. $\Box$

In Examples (iii), (iv), (vi), (vii), and (ix), the multiplication is commutative, i.e.,

$$\alpha\beta = \beta\alpha \quad \text{for all } \alpha, \beta \in G.$$

In Examples (i) and (ii), as we have seen, it is generally not. A group is called *commutative* or *abelian* if the group operation is commutative. Group elements can be formally manipulated in the ways you expect:

THEOREM 5.1
Let G be a group.

(i) *The identity element is unique.*

(ii) *The inverse of an element α is unique.*

(iii) *For any $\alpha \in G$, $\left(\alpha^{-1}\right)^{-1} = \alpha$.*

(iv) *For any $\alpha, \beta \in G$, $(\alpha\beta)^{-1} = \beta^{-1}\alpha^{-1}$.*

(v) *Cancellation rules: For any $\alpha, \beta, \gamma \in G$,*

(a) *if $\alpha\beta = \alpha\gamma$, then $\beta = \gamma$;*
(b) *if $\beta\alpha = \gamma\alpha$, then $\beta = \gamma$.*

PROOF   (i) Suppose $\epsilon \in G$ is another identity element, i.e.,

$$\epsilon\alpha = \alpha\epsilon = \alpha, \text{ for all } \alpha \in G.$$

Take $\alpha = 1$. Thus $\epsilon \cdot 1 = 1$. But since 1 is also an identity element $\epsilon \cdot 1 = \epsilon$. Therefore $\epsilon = 1$.
(ii) Let α' be another inverse for α. So

$$1 = \alpha\alpha' = \alpha'\alpha.$$

Multiply the first equation on the left by α^{-1} and use associativity:

$$\alpha^{-1} = \alpha^{-1}(\alpha\alpha') = (\alpha^{-1}\alpha)\alpha' = 1 \cdot \alpha' = \alpha'.$$

Thus $\alpha^{-1} = \alpha'$.
(iii) The inverse of α^{-1} is characterized by the property

$$\left(\alpha^{-1}\right)^{-1}\alpha^{-1} = 1 = \alpha^{-1}\left(\alpha^{-1}\right)^{-1}.$$

But comparing with property (iii) in the definition of a group (Definition 5.2), we see that α also satisfies these equations. Since there is a unique inverse for α^{-1}, we must have $\left(\alpha^{-1}\right)^{-1} = \alpha$.

(iv) We have

$$(\beta^{-1}\alpha^{-1})(\alpha\beta) = \beta^{-1}(\alpha^{-1}\alpha)\beta = (\beta^{-1} \cdot 1)\beta = \beta^{-1}\beta = 1 \ .$$

Similarly

$$(\alpha\beta)(\beta^{-1}\alpha^{-1}) = 1 \ .$$

Therefore $\beta^{-1}\alpha^{-1}$ is the inverse of $\alpha\beta$.

(v) Suppose

$$\alpha\beta = \alpha\gamma \ .$$

Multiply on the left by α^{-1}:

$$\alpha^{-1}(\alpha\beta) = \alpha^{-1}(\alpha\gamma) \ .$$

By associativity,

$$(\alpha^{-1}\alpha)\beta = (\alpha^{-1}\alpha)\gamma \ ,$$

which gives

$$1 \cdot \beta = 1 \cdot \gamma \quad \text{or} \quad \beta = \gamma \ .$$

The proof of (b) is analogous.   ∎

If a group G is finite, then the number of elements in G is the *order* of G, written $|G|$. If G is infinite, we write $|G| = \infty$. Suppose $\alpha \in G$ and there exists $n \in \mathbb{N}$ so that $\alpha^n = 1$. Then the *order* of α, written $|\alpha|$, is the smallest such n. If no such n exists, we say that α has infinite order, $|\alpha| = \infty$. As in Chapters 3 and 4, we can consider $\langle\alpha\rangle = \{\alpha^n \mid n \in \mathbb{Z}\}$. It is easy to see that $\langle\alpha\rangle$ with the operation inherited from G is a group. And

$$|\langle\alpha\rangle| = |\alpha| \ .$$

Of course these definitions just generalize the definitions presented for permutation groups and linear groups.

Example 5.2

In Chapter 1, following Remark 1.1, we defined $\varphi(n)$ to be the number of integers $a, 0 < a \le n$, which are relatively prime to n. So the order of $(\mathbb{Z}/m\mathbb{Z})^\times$ is $\varphi(m)$. If $(10, m) = 1$, then $\bar{10} \in (\mathbb{Z}/m\mathbb{Z})^\times$. By Theorem 1.3 the order of $\bar{10}$ is the length of the period of $1/m$.   □

Homomorphisms

In Example (xiii), there is a natural mapping from B_n to S_n. Each braid ζ defines a permutation α_ζ with

$$\alpha_\zeta(j) = \text{end point of strand starting at } j \ .$$

For any $\alpha \in S_n$ there are many braids $\zeta \in B_n$ with $\alpha_\zeta = \alpha$. So this mapping is surjective but not injective. It is well-behaved with respect to the multiplications in B_n and S_n: from the diagrams we see that

$$\alpha_{\zeta_2 \zeta_1} = \alpha_{\zeta_2} \alpha_{\zeta_1}$$

for any $\zeta_1, \zeta_2 \in B_n$. Since we know a bit about S_n, the mapping tells us something about what the group B_n looks like. Such a mapping from one group to another is called a *homomorphism*.

DEFINITION 5.3   *Let G and H be groups. A mapping $f : G \to H$ is called a (group) homomorphism if for any $\alpha, \beta \in G$,*

$$f(\alpha\beta) = f(\alpha)f(\beta) .$$

Examples 5.3

We have already seen several other mappings between groups that are homomorphisms. At the end of Chapter 2 we defined the permutation matrix $I_\alpha \in GL(n, \mathbb{R})$ belonging to a permutation $\alpha \in S_n$. Recall that the jth row of I_α is just the $\alpha(j)$th row of the identity matrix. We saw that

$$I_{\alpha\beta} = I_\alpha I_\beta$$

for any $\alpha, \beta \in S_n$. Thus the mapping $p : S_n \to GL(n, \mathbb{R})$ given by

$$p : \alpha \mapsto I_\alpha$$

is a homomorphism. We also know that for any field F, the determinant

$$\det : GL(n, F) \to F^\times$$

satisfies

$$\det(\alpha\beta) = \det(\alpha) \det(\beta)$$

for any $\alpha, \beta \in GL(n, F)$. So det is also a homomorphism. A final example: addition in $\mathbb{Z}/n\mathbb{Z}$ was defined so that the canonical mapping $\mathbb{Z} \to \mathbb{Z}/n\mathbb{Z}$ given by $a \mapsto \bar{a}$ is a homomorphism. ▯

Homomorphisms have the formal properties that you would expect.

THEOREM 5.2

Let $f : G \to H$ be a group homomorphism. Then

(i) $f(1_G) = 1_H$;

(ii) $f(\alpha^{-1}) = f(\alpha)^{-1}$ for any $\alpha \in G$;

(iii) if f is bijective, then $f^{-1} : H \to G$ is also a homomorphism.

PROOF   We have
$$f(1_G) = f(1_G 1_G) = f(1_G)^2 .$$

Multiply on the left by $f(1_G)^{-1}$:

$$1_H = f(1_G)^{-1} f(1_G) = f(1_G)^{-1} f(1_G)^2 = f(1_G) .$$

It follows that for any $\alpha \in G$,

$$1_H = f(1_G) = f(\alpha \alpha^{-1}) = f(\alpha) f(\alpha^{-1}) .$$

Therefore $f(\alpha^{-1}) = f(\alpha)^{-1}$. Lastly suppose that f is bijective. For any $\sigma, \tau \in H$, we have

$$f\big(f^{-1}(\sigma \tau)\big) = \sigma \tau .$$

On the other hand,

$$f\big(f^{-1}(\sigma) f^{-1}(\tau)\big) = f\big(f^{-1}(\sigma)\big) f\big(f^{-1}(\tau)\big) = \sigma \tau$$

as well. Since f is injective, it follows that

$$f^{-1}(\sigma \tau) = f^{-1}(\sigma) f^{-1}(\tau) . \quad \blacksquare$$

A homomorphism that is bijective is called an *isomorphism.* Two groups G and H are said to be *isomorphic,* written $G \cong H$, if there exists an isomorphism $f : G \to H$.

Example 5.4
Let $\mathbb{R}^+$ denote the positive real numbers. Under multiplication $\mathbb{R}^+$ is an abelian group. Let

$$f : \mathbb{R} \to \mathbb{R}^+$$

be the exponential mapping

$$f : x \mapsto e^x .$$

Since $e^{x+y} = e^x e^y$, f is a homomorphism. It has an inverse, the logarithm:

$$\log : \mathbb{R}^+ \to \mathbb{R} .$$

So the exponential is an isomorphism of groups. The formula $\log(zw) = \log(z) + \log(w)$ just says that log is a homomorphism.   □

Given two groups G and H, we define their *direct product* to be the set $G \times H$ with the operation:

$$\big((\alpha_1, \beta_1), (\alpha_2, \beta_2)\big) \mapsto (\alpha_1 \alpha_2, \beta_1 \beta_2)$$

where $\alpha_1, \alpha_2 \in G$ and $\beta_1, \beta_2 \in H$. It is not hard to see that this operation makes $G \times H$ into a group (see Exercise 5.16). For example, $\mathbb{R}^2$ with vector addition is the direct product of $\mathbb{R}$ with itself. The two projections:

$$p_1 : G \times H \to G$$
$$p_1 : (\alpha, \beta) \mapsto \alpha$$

and

$$p_2 : G \times H \to H$$
$$p_2 : (\alpha, \beta) \mapsto \beta \, ,$$

for $\alpha \in G$, $\beta \in H$, are homomorphisms.

Example 5.5

(i) Let us show that $GL(3, \mathbb{R})$ is isomorphic to $SL(3, \mathbb{R}) \times \mathbb{R}^\times$. We define a mapping

$$f : SL(3, \mathbb{R}) \times \mathbb{R}^\times \to GL(3, \mathbb{R})$$

by

$$f : (\alpha, a) \mapsto a\alpha$$

for $\alpha \in SL(3, \mathbb{R})$ and $a \in \mathbb{R}^\times$. We see immediately that f is a homomorphism. To define an inverse to f, first notice that the mapping

$$h : \mathbb{R}^\times \to \mathbb{R}^\times$$

given by

$$h(a) = a^3$$

is an isomorphism. And if $\alpha \in GL(3, \mathbb{R})$, then $(\det \alpha)^{-1/3}\alpha$ has determinant 1. Therefore, since det is a homomorphism, the mapping

$$g : \alpha \mapsto \left((\det \alpha)^{-1/3}\alpha, (\det \alpha)^{1/3}\right)$$

is a homomorphism from $GL(3, \mathbb{R})$ to $SL(3, \mathbb{R}) \times \mathbb{R}^\times$. It is clearly inverse to f.

(ii) Suppose that $m, n \in \mathbb{N}$ and $(m, n) = 1$. For any $a \in \mathbb{Z}$, let $\bar{a}$ denote its residue class mod mn, $\bar{a}_1$ denote its residue class mod m, and $\bar{a}_2$ denote its residue class mod n. Define a mapping

$$g : \mathbb{Z}/mn\mathbb{Z} \to (\mathbb{Z}/m\mathbb{Z}) \times (\mathbb{Z}/n\mathbb{Z})$$

by

$$g(\bar{a}) := (\bar{a}_1, \bar{a}_2) \, .$$

This mapping is well-defined because for any $k \in \mathbb{Z}$,

$$a + kmn \equiv a \pmod{m} \qquad \text{and} \qquad a + kmn \equiv a \pmod{n} .$$

We have

$$g(\bar{a} + \bar{b}) = (\bar{a}_1 + \bar{b}_1, \bar{a}_2 + \bar{b}_2) = (\bar{a}_1, \bar{a}_2) + (\bar{b}_1, \bar{b}_2) = g(\bar{a}) + g(\bar{b}) .$$

So g is a homomorphism. Now the *Chinese Remainder Theorem* (Theorem 1.5) says precisely that g is an isomorphism. Why is this? Well, what does it mean for g to be surjective? Given integers a and b, we must show that there exists an integer c such that

$$c \equiv a \pmod{m} \quad \text{and} \quad c \equiv b \pmod{n} .$$

To say that g is injective is to say that such an integer c is unique mod mn. $\quad \square$

Exercises

5.1 Which of the following groups are abelian: the permutation group V, the linear groups T, $F(p)$, or $G(p)$?

5.2 Let G be a group. For $\alpha \in G$, let α^n be the n-fold product of α with itself, for $n > 0$, and the n-fold product of α^{-1} with itself, for $n < 0$. Show that

$$\alpha^m \alpha^n = \alpha^{m+n}$$
$$(\alpha^m)^n = \alpha^{mn}$$

for any $m, n \in \mathbb{Z}$.

5.3 What is the order of $(\mathbb{Z}/16\mathbb{Z})^\times$? $(\mathbb{Z}/24\mathbb{Z})^\times$? For each group make a table that gives the order of the elements in the group.

5.4 What is the order of any element in the additive group of $\mathbb{F}_p$? What are the orders of the elements in the multiplicative group $\mathbb{F}_{11}^\times$?

5.5 Let α be an element of order n in a group G. Suppose that $\alpha^r = 1$. Show that $n \mid r$.

5.6 • Suppose that α is an element of order n in a group G. For any $m \in \mathbb{N}$, prove that
$$|\alpha^m| = n/(m, n) .$$

5.7 • Let α and β be elements of prime order in a group G, with $|\alpha| \neq |\beta|$. Suppose that $\alpha\beta = \beta\alpha$. Prove that

$$|\alpha\beta| = |\alpha||\beta| .$$

5.8 • Let α and β be elements of finite order in a group G, with $(|\alpha|, |\beta|) = 1$. Suppose that $\alpha\beta = \beta\alpha$. Prove that

$$|\alpha\beta| = |\alpha||\beta| .$$

5.9 Determine the structure of the braid group B_2.

5.10 Show that an isometry is injective.

5.11 • Let exp : $\mathbb{R} \to S$ be the exponential mapping

$$\exp(x) = e^{2\pi ix} .$$

Verify that exp is a homomorphism. Is exp injective? Surjective?

5.12 Show that the mapping

$$h : \mathbb{R}^\times \to \mathbb{R}^\times$$

given by

$$h(a) = a^3$$

is an isomorphism.

5.13 Check that the mapping in Example 4.1(i) is an isomorphism.

5.14 Given a natural number n, prove that if G is an abelian group, then the mapping $f : G \to G$, where

$$f : \alpha \mapsto \alpha^n ,$$

is a homomorphism.

5.15 Prove that the permutation groups V and V' are isomorphic.

5.16 Verify that the direct product of two groups with the operation given is a group.

5.17 Give an example of two finite groups of the same order that are not isomorphic.

5.18 (a) Show that

$$(\mathbb{Z}/16\mathbb{Z})^\times \cong (\mathbb{Z}/2\mathbb{Z}) \times (\mathbb{Z}/4\mathbb{Z}) .$$

(b) Show that

$$(\mathbb{Z}/24\mathbb{Z})^\times \cong (\mathbb{Z}/2\mathbb{Z})^3 .$$

5.19 Let $f : G \to H$ be a homomorphism of groups and let $g \subset G$ be a set of generators of G. Suppose that

$$h := \{f(\alpha) \mid \alpha \in g\}$$

generates H. Prove that f is surjective.

5.20 • Suppose that G and H are finite groups with the same order. Let $f : G \to H$ be a homomorphism. Show that

    (a) if f is surjective, then it is injective;

    (b) if f is injective, then it is surjective.

5.21 • Suppose G is a group. An isomorphism $f : G \to G$ is called an *automorphism*. Let $\mathrm{Aut}(G)$ denote the set of automorphisms of G. Prove that $\mathrm{Aut}(G)$ is a group under composition of mappings.

5.22 Prove that $\mathrm{Aut}(\mathbb{Z}/n\mathbb{Z}) \cong (\mathbb{Z}/n\mathbb{Z})^{\times}$.

5.23 Suppose that $m, n \in \mathbb{N}$ and $(m, n) = 1$. Define a mapping

$$h : (\mathbb{Z}/m\mathbb{Z}) \times (\mathbb{Z}/n\mathbb{Z}) \to \mathbb{Z}/mn\mathbb{Z}$$

by

$$h : (\bar{a}, \bar{b}) \mapsto \overline{na + mb}\,.$$

Show that h is well-defined and is a homomorphism. Prove that h is an isomorphism.

5.24 • Recall that the order of $(\mathbb{Z}/m\mathbb{Z})^{\times}$ is $\varphi(m)$ (see Example 5.2).

    (a) Suppose that $(m, n) = 1$. Prove that

$$g : (\mathbb{Z}/mn\mathbb{Z})^{\times} \xrightarrow{\ \cong\ } (\mathbb{Z}/m\mathbb{Z})^{\times} \times (\mathbb{Z}/n\mathbb{Z})^{\times}\,,$$

    where g is the mapping defined in Example 5.5(ii).

    (b) Show that $\varphi(mn) = \varphi(m)\varphi(n)$.

    (c) Suppose that

$$n = p_1^{j_1} \cdots p_r^{j_r}\,,$$

    where $p_1, \dots, p_r$ are distinct primes and $j_1, \dots, j_r > 0$. Show that

$$\varphi(n) = (p_1^{j_1} - p_1^{j_1 - 1}) \cdots (p_r^{j_r} - p_r^{j_r - 1}) = n(1 - 1/p_1) \cdots (1 - 1/p_r)$$

    (see Exercise 1.11).

Chapter 6

Subgroups

Definition

Subgroups are subsets of groups, which are groups themselves under the operation inherited from the group. Of course for this to be possible, the product of two elements of the subgroup must lie within it, and so must the inverse of every element. It turns out that this is in fact enough.

DEFINITION 6.1   *Let G be a group. A nonempty subset $H \subseteq G$ is a subgroup of G, written $H < G$, if for all $\alpha, \beta \in H$*

$$\alpha\beta \in H \tag{i}$$

$$\alpha^{-1} \in H . \tag{ii}$$

Condition (i) ensures that the group operation on G gives us an operation on H. It is associative because the operation on G is. Since H is nonempty, there exists an element $\alpha \in H$. By (ii), $\alpha^{-1} \in H$ as well. Therefore $1 = \alpha\alpha^{-1} \in H$ by (i). And by (ii) again, the inverse of every element in H lies in H. So H, with the operation inherited from G, is a group.

Looking back at Chapters 3 and 4 we see that permutation groups were defined as subgroups of S_n and linear groups as subgroups of $GL(n, F)$. Looking at the list of examples in Chapter 5, we see that in Example 5.1

$$\mathbb{Z} < \mathbb{Q} < \mathbb{R} < \mathbb{C} .$$

Orthogonal Groups

There are some more examples of linear groups that are of interest. These are the orthogonal groups, which consist of those isometries that are linear mappings. Recall that a real $n \times n$ matrix α is *orthogonal* if

$$\|\alpha v\| = \|v\|$$

for all $v \in \mathbb{R}^n$. If we denote by $O(n)$ the set of orthogonal matrices, then

$$O(n) = GL(n, \mathbb{R}) \cap Iso(n) .$$

Thus $O(n)$ is a subgroup of $Iso(n)$ and of $GL(n, \mathbb{R})$. The special orthogonal group, $SO(n)$, is given by

$$SO(n) = O(n) \cap SL(n, \mathbb{R}) .$$

So it is a subgroup of $O(n)$ and $SL(n, \mathbb{R})$ (as well as of $Iso(n)$ and of $GL(n, \mathbb{R})$).

It is not hard to describe the elements of $SO(2)$. A matrix $\left(\begin{smallmatrix} a & b \\ c & d \end{smallmatrix} \right) \in SO(2)$ if and only if

$$a^2 + c^2 = 1$$
$$b^2 + d^2 = 1$$
$$ab + cd = 0$$
$$ad - bc = 1 .$$

The first equation tells us that there exists $t \in \mathbb{R}$ such that $(a, c) = (\cos t, \sin t)$. The third equation implies that $(b, d) = r(-\sin t, \cos t)$ for some $r \in \mathbb{R}$. From the second, we can conclude that $r = \pm 1$, and from the fourth, that $r = 1$. Thus

$$\begin{pmatrix} a & b \\ c & d \end{pmatrix} = \begin{pmatrix} \cos t & -\sin t \\ \sin t & \cos t \end{pmatrix} .$$

So the elements of $SO(2)$ are just the rotations about the origin and t is the angle of rotation. Multiplying two matrices corresponds to adding the angles of rotation.

We can also show what elements of $O(2)$, whose determinant is -1, look like. Set

$$\rho = \begin{pmatrix} -1 & 0 \\ 0 & 1 \end{pmatrix} .$$

If $\alpha \in O(2)$ and $\det \alpha = -1$, then $\det \rho\alpha = 1$. So $\rho\alpha$ is of the form

$$\begin{pmatrix} \cos t & -\sin t \\ \sin t & \cos t \end{pmatrix}$$

for some t, and therefore α can be written

$$\alpha = \begin{pmatrix} -\cos t & \sin t \\ \sin t & \cos t \end{pmatrix} .$$

Viewed geometrically ρ is a reflection in the line $x = 0$. It is easy to say what a reflection is in general, in any dimension. Let $v \in \mathbb{R}^n$ be a unit vector, and set $W = v^\perp$. Define $\rho_v \in O(n)$ by

$$\rho_v(v) = -v \quad , \qquad \rho_{v|W} = I_W .$$

Then ρ_v is a reflection in the hyperplane W. For an arbitrary vector $w \in \mathbb{R}^n$ we have

$$\rho_v(w) = w - 2(w \cdot v)v .$$

If we pick an orthonormal basis $\{v_2, \ldots, v_n\}$ of W, the matrix of ρ_v with respect to $\{v, v_2, \ldots, v_n\}$ is

$$\begin{pmatrix} -1 & 0 & \ldots & 0 \\ 0 & 1 & \ldots & 0 \\ \vdots & \vdots & \ddots & \vdots \\ 0 & 0 & \ldots & 1 \end{pmatrix} .$$

Thus $\det \rho_v = -1$.

Lastly let us describe what an element $\alpha \in SO(3)$ looks like. First we check that 1 is an eigenvalue: since $\det \alpha = 1$,

$$\begin{aligned} \det(\alpha - I) &= \det(\alpha^t - I) = \det \alpha \det(\alpha^t - I) \\ &= \det(\alpha \alpha^t - \alpha) = \det(I - \alpha) \\ &= -\det(\alpha - I) . \end{aligned}$$

Therefore $\det(\alpha - I) = 0$. Let v then be a unit eigenvector with eigenvalue 1. Since α is orthogonal, the plane $W := v^\perp$ is invariant under α and it is easy to see that

$$\det \alpha_{|W} = \det \alpha = 1 .$$

So $\alpha_{|W} \in SO(2)$. Now let $\{w_1, w_2\}$ be an orthonormal basis of W. Then with respect to the orthonormal basis $\{v, w_1, w_2\}$ of $\mathbb{R}^3$, a matrix of α has the form

$$\begin{pmatrix} 1 & 0 & 0 \\ 0 & \cos t & -\sin t \\ 0 & \sin t & \cos t \end{pmatrix} .$$

In other words, α is a rotation about the line through v through the angle t. Notice that

$$\operatorname{tr} \alpha = 2 \cos t .$$

Cyclic Subgroups and Generators

If G is a group and $\alpha \in G$ then $\langle \alpha \rangle = \{\alpha^n \mid n \in \mathbb{Z}\}$ is called the *cyclic* subgroup generated by α. If $\langle \alpha \rangle = G$, then G is a *cyclic* group.

THEOREM 6.1
An infinite cyclic group is isomorphic to $\mathbb{Z}$. A cyclic group of order n is isomorphic to $\mathbb{Z}/n\mathbb{Z}$.

PROOF   Let G be a cyclic group with generator α. Define a mapping $f : \mathbb{Z} \to G$ by

$$f : m \mapsto \alpha^m$$

for $m \in \mathbb{Z}$. Since

$$f(m_1 + m_2) = \alpha^{m_1 + m_2} = \alpha^{m_1} \alpha^{m_2} = f(m_1) f(m_2) \, ,$$

f is a homomorphism. Since $G = \langle \alpha \rangle$, f is surjective.

Now there are two possibilities. First suppose that $|G| = \infty$. Then $\alpha^m \neq 1$ for all $m \neq 0$. If $\alpha^{m_1} = \alpha^{m_2}$ for some m_1, m_2, then $\alpha^{m_1 - m_2} = 1$. But then $m_1 - m_2 = 0$. So f is injective and therefore is an isomorphism.

The second possibility is that $|G| = n$, for some $n \in \mathbb{N}$, i.e., $|\alpha| = n$. Now $f(m + sn) = \alpha^{m+sn} = \alpha^m = f(m)$ for any s. So f defines a mapping

$$\bar{f} : \mathbb{Z}/n\mathbb{Z} \to G$$

which is also a surjective homomorphism. Since both groups have order n, Exercise 5.20 implies that f is also injective and therefore an isomorphism.   ∎

Thus any two cyclic groups of the same order are isomorphic. It is not true in general that two groups of the same order are isomorphic. For example, the group V, which has order 4, is not isomorphic to a cyclic group of order 4 because it has no element of order 4.

More generally, if $g \subset G$, then the subgroup generated by g, written $\langle g \rangle$, is the subgroup consisting of all elements of G that can be expressed in terms of the elements of g and their inverses. We have to make this more precise. If G is finite, then we have

$$\langle g \rangle = \bigcup_{i=0}^{\infty} g^i$$

as in Chapter 3. Otherwise the description is more complicated. We must include expressions of the form

$$\alpha_1^{e_1} \alpha_2^{e_2} \cdots \alpha_k^{e_k}$$

where $e_i = \pm 1$ for all i and $\alpha_i \in g$. Such expressions are called *words* in g.

THEOREM 6.2
Let

$$H = \left\{ \alpha_1^{e_1} \alpha_2^{e_2} \cdots \alpha_k^{e_k} \mid k > 0, e_i = \pm 1, \alpha_i \in g, 1 \leq i \leq k \right\} \, .$$

Then H is the smallest subgroup of G containing g and is the intersection of all the subgroups of G containing g.

PROOF   Certainly $H \supset g$ and is thus nonempty. The product of any two words in g is another word in g. And the inverse of a word is as well. So H is a subgroup. Clearly

H is contained in any subgroup of G that contains g. Therefore H is the smallest such subgroup. To see that H is the intersection of all the subgroups containing g we need only show:

LEMMA 6.1
The intersection K of a collection C of subgroups of G is again a subgroup.

Well, $K \neq \emptyset$ since $1 \in K$. If $\alpha, \beta \in K$, then for any $L \in C$ we have $\alpha, \beta \in L$ and therefore $\alpha\beta \in L$ and $\alpha^{-1} \in L$. It follows that $\alpha\beta, \alpha^{-1} \in K$. And thus K is a subgroup of G. $\blacksquare$

Example 6.1
In the group $GL(n, \mathbb{Z})$ we have the subgroup

$$SL(n, \mathbb{Z}) := \{\alpha \in GL(n, \mathbb{Z}) \mid \det \alpha = 1\} \,.$$

The purpose of this example is to find a pair of generators for $SL(2, \mathbb{Z})$. Suppose

$$\alpha = \begin{pmatrix} a & b \\ c & d \end{pmatrix} \in SL(2, \mathbb{Z}) \,.$$

By applying the Euclidean algorithm to a and b, we are going to find an expression for α in terms of

$$\sigma = \begin{pmatrix} 1 & 1 \\ 0 & 1 \end{pmatrix} \quad \text{and} \quad \tau = \begin{pmatrix} 1 & 0 \\ 1 & 1 \end{pmatrix} \,.$$

In particular, this will prove that

$$SL(2, \mathbb{Z}) = \left\langle \begin{pmatrix} 1 & 1 \\ 0 & 1 \end{pmatrix}, \begin{pmatrix} 1 & 0 \\ 1 & 1 \end{pmatrix} \right\rangle \,.$$

Now applying the Euclidean algorithm to a and b gives us a list of equations:

$$
\begin{array}{ll}
a = qb + r & 0 \le r < b \\
b = q_1 r + r_1 & 0 \le r_1 < r \\
\vdots & \vdots \\
r_{i-1} = q_{i+1} r_i + r_{i+1} & 0 \le r_{i+1} < r_i \\
\vdots & \vdots \\
r_{n-2} = q_n r_{n-1} + r_n & 0 \le r_n < r_{n-1} \\
r_{n-1} = q_{n+1} r_n &
\end{array}
$$

Since $ad - bc = 1$, it follows that $(a, b) = 1$ and therefore $r_n = 1$. We begin by multiplying α on the right by τ^{-q}:

$$\begin{pmatrix} a & b \\ c & d \end{pmatrix} \begin{pmatrix} 1 & 0 \\ -q & 1 \end{pmatrix} = \begin{pmatrix} a - bq & b \\ * & * \end{pmatrix} = \begin{pmatrix} r & b \\ * & * \end{pmatrix}$$

where the entries * denote integers which are not important to us, and the resulting matrix has determinant 1. Next we multiply it by σ^{-q_1}:

$$\begin{pmatrix} r & b \\ * & * \end{pmatrix} \begin{pmatrix} 1 & -q_1 \\ 0 & 1 \end{pmatrix} = \begin{pmatrix} r & b - q_1 r \\ * & * \end{pmatrix} = \begin{pmatrix} r & r_1 \\ * & * \end{pmatrix} .$$

Again the resulting matrix has determinant 1. In general we will have a matrix in $SL(2, \mathbb{Z})$ of the form

$$\begin{pmatrix} r_{i-1} & r_i \\ * & * \end{pmatrix} \quad \text{or} \quad \begin{pmatrix} r_i & r_{i-1} \\ * & * \end{pmatrix} ,$$

depending upon whether i is odd or even. We multiply it by $\tau^{-q_{i+1}}$, respectively $\sigma^{-q_{i+1}}$:

$$\begin{pmatrix} r_{i-1} & r_i \\ * & * \end{pmatrix} \begin{pmatrix} 1 & 0 \\ -q_{i+1} & 1 \end{pmatrix} = \begin{pmatrix} r_{i+1} & r_i \\ * & * \end{pmatrix}$$

$$\begin{pmatrix} r_i & r_{i-1} \\ * & * \end{pmatrix} \begin{pmatrix} 1 & -q_{i+1} \\ 0 & 1 \end{pmatrix} = \begin{pmatrix} r_i & r_{i+1} \\ * & * \end{pmatrix} .$$

At the end we are left with

$$\begin{pmatrix} r_n & 0 \\ * & * \end{pmatrix} \quad \text{or} \quad \begin{pmatrix} 0 & r_n \\ * & * \end{pmatrix} .$$

Now $r_n = 1$ and the matrices have determinant 1. So in the first case we have a matrix of the form

$$\begin{pmatrix} 1 & 0 \\ k & 1 \end{pmatrix} = \tau^k ,$$

for some integer k. In the second case we have a matrix of the form

$$\begin{pmatrix} 0 & 1 \\ -1 & k \end{pmatrix}$$

for some $k \in \mathbb{Z}$. And

$$\begin{pmatrix} 0 & 1 \\ -1 & k \end{pmatrix} = \begin{pmatrix} 1 & 0 \\ k & 1 \end{pmatrix} \begin{pmatrix} 0 & 1 \\ -1 & 0 \end{pmatrix} = \begin{pmatrix} 1 & 0 \\ k & 1 \end{pmatrix} \begin{pmatrix} 1 & 1 \\ 0 & 1 \end{pmatrix} \begin{pmatrix} 1 & 0 \\ -1 & 1 \end{pmatrix} \begin{pmatrix} 1 & 1 \\ 0 & 1 \end{pmatrix} = \tau^k \sigma \tau^{-1} \sigma .$$

So we have shown that

$$\alpha \tau^{-q} \sigma^{-q_1} \cdots = \tau^k \text{ or } \tau^k \sigma \tau^{-1} \sigma ,$$

which means that

$$\alpha = \tau^k \cdots \sigma^{q_1} \tau^q \text{ or } \tau^k \sigma \tau^{-1} \sigma \cdots \sigma^{q_1} \tau^q .$$

Let us do a numerical example. Take

$$\alpha = \begin{pmatrix} 19 & 7 \\ 27 & 10 \end{pmatrix} .$$

Applying the Euclidean algorithm gives us

$$19 = 2 \cdot 7 + 5$$
$$7 = 5 + 2$$
$$5 = 2 \cdot 2 + 1$$
$$2 = 2 \cdot 1 \, .$$

So we first multiply α by τ^{-2}:

$$\begin{pmatrix} 19 & 7 \\ 27 & 10 \end{pmatrix} \begin{pmatrix} 1 & 0 \\ -2 & 1 \end{pmatrix} = \begin{pmatrix} 5 & 7 \\ 7 & 10 \end{pmatrix} \, .$$

Then we multiply the result by σ^{-1}:

$$\begin{pmatrix} 5 & 7 \\ 7 & 10 \end{pmatrix} \begin{pmatrix} 1 & -1 \\ 0 & 1 \end{pmatrix} = \begin{pmatrix} 5 & 2 \\ 7 & 3 \end{pmatrix} \, .$$

Next we multiply again by τ^{-2}:

$$\begin{pmatrix} 5 & 2 \\ 7 & 3 \end{pmatrix} \begin{pmatrix} 1 & 0 \\ -2 & 1 \end{pmatrix} = \begin{pmatrix} 1 & 2 \\ 1 & 3 \end{pmatrix} \, .$$

Lastly we multiply by σ^{-2}:

$$\begin{pmatrix} 1 & 2 \\ 1 & 3 \end{pmatrix} \begin{pmatrix} 1 & -2 \\ 0 & 1 \end{pmatrix} = \begin{pmatrix} 1 & 0 \\ 1 & 1 \end{pmatrix} \, .$$

Thus

$$\alpha \tau^{-2} \sigma^{-1} \tau^{-2} \sigma^{-2} = \tau \, ,$$

or

$$\alpha = \tau \sigma^2 \tau^2 \sigma \tau^2 \, . \quad \square$$

Example 6.2

It is not hard to describe a set of generators of the braid group B_n. For $1 \leq i < n$, let ζ_i be the following braid: if $j \neq i, i + 1$, the jth strand joins point j on the first line to point j on the second without crossing any other strand. The ith strand joins point i on the first line to point $i + 1$ on the second, crossing under the $(i + 1)$st strand, which joins point $i + 1$ on the first to point i on the second.

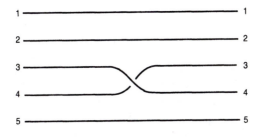

To see that $\{\zeta_1, \dots, \zeta_{n-1}\}$ generates B_n, take any braid ζ, and perturb the strands so that no two crossovers occur at the same level. Draw lines parallel to the two lines through end points, which subdivide the region between them into slices, each containing only one crossover. Then in each slice, the braid is one of the $\zeta_i^{\pm 1}$.

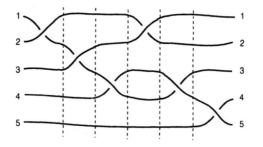

For example, the braid above can be written:

$$\zeta_1^{-1}\zeta_2^{-1}\zeta_3\zeta_1^{-1}\zeta_3\zeta_4 \ .$$

If we let

$$\rho = \zeta_1\zeta_2\cdots\zeta_{n-1} \ ,$$

then one can show that $\{\rho, \zeta_1\}$ also generates B_n.   ▯

Kernel and Image of a Homomorphism

Suppose G and H are groups and $f : G \to H$ is a homomorphism. The *kernel* of f, written $\ker(f)$, is defined by

$$\ker(f) := \{\alpha \in G \mid f(\alpha) = 1_H\} \subset G \ .$$

It measures how close f is to being injective, as we shall see below. The *image* of f, written $\mathrm{im}(f)$, is the set of images of the elements of G in H, i.e.,

$$\mathrm{im}(f) := \{f(\alpha) \mid \alpha \in G\} \subset H \ .$$

THEOREM 6.3
The kernel of a homomorphism f is a subgroup of G, and the image a subgroup of H. A homomorphism is injective if and only if its kernel is trivial.

We just need to verify that $\ker(f)$ and $\mathrm{im}(f)$ fulfill the definition of a subgroup. Suppose $\alpha, \beta \in G$, $f(\alpha) = 1$ and $f(\beta) = 1$. Then

$$f(\alpha\beta) = f(\alpha)f(\beta) = 1$$

and

$$f(\alpha^{-1}) = f(\alpha)^{-1} = 1 \, .$$

Therefore $\ker(f)$ is a subgroup of G. And for any $\alpha, \beta \in G$, $f(\alpha)f(\beta) = f(\alpha\beta) \in \operatorname{im}(f)$, and $f(\alpha)^{-1} = f(\alpha^{-1}) \in \operatorname{im}(f)$. Thus $\operatorname{im}(f)$ is a subgroup of H.

Lastly, suppose f is injective. Then in particular, if $f(\alpha) = 1$ for some $\alpha \in G$, since $f(1) = 1$, we have $\alpha = 1$. Thus $\ker(f) = \{1\}$. On the other hand, assume $\ker(f) = \{1\}$. Now if $f(\alpha) = f(\beta)$ for some $\alpha, \beta \in G$, we have

$$1 = f(\alpha)f(\beta)^{-1} = f(\alpha)f(\beta^{-1}) = f(\alpha\beta^{-1}) \, .$$

So $\alpha\beta^{-1} \in \ker(f)$, and therefore $\alpha\beta^{-1} = 1$, i.e., $\alpha = \beta$. Thus f is injective.

Examples 6.3

(i) Recall that $\det : GL(n, F) \to F^\times$ is a homomorphism for any field F. By the definition of $SL(n, F)$

$$\ker(\det) = SL(n, F) \, .$$

(ii) The kernel of the canonical map: $\mathbb{Z} \to \mathbb{Z}/n\mathbb{Z}$ is the subgroup $n\mathbb{Z} \subset \mathbb{Z}$.

(iii) Let $f : S \to S$ be given by $f(z) := z^n$, for some $n \in \mathbb{N}$. The kernel of f is

$$\left\{ z \in \mathbb{C} \mid z^n = 1 \right\} \, .$$

A complex number z satisfying this equation is called an nth root of unity. We shall denote the group of nth roots of unity by μ_n.

(iv) If G and H are two groups, then the kernel of the projection $p_1 : G \times H \to G$ is H.

(v) In Chapter 5, we saw that there is a natural homomorphism

$$br : B_n \to S_n \, , \quad \text{where } br(\zeta) = \alpha_\zeta \, .$$

The permutation α_ζ maps i to the end point of the ith strand of ζ. So

$$\alpha_{\zeta_i} = (i \; i+1) \, .$$

Since

$$\{(1 \; 2), (2 \; 3), \; \ldots \, , \; (n-1 \; n)\}$$

generates S_n, the homomorphism br is surjective. □

Exercises

6.1 What are the elements of finite order in $SO(2)$?

6.2 • Show that μ_n is a cyclic subgroup of S. Notice that every element of finite order of S is an nth root of unity for some n.

6.3 Use the algorithm in Example 6.1 to write $\begin{pmatrix} 7 & 31 \\ 2 & 9 \end{pmatrix}$ as a word in σ and τ.

6.4 Write a *Mathematica* function that expresses an element of $SL(2, \mathbb{Z})$ as a word in σ and τ.

6.5 Let $a, b \in \mathbb{Z}$ be relatively prime. The Euclidean algorithm produces integers s and t such that

$$sa - tb = 1 .$$

Set

$$\alpha = \begin{pmatrix} a & b \\ t & s \end{pmatrix} .$$

Prove that if one applies the algorithm in example 6.1 to α, then the integer k there is 0 (cf. Exercise 1.13).

6.6 Show that $\mathbb{F}_5^\times$, $\mathbb{F}_7^\times$, and $\mathbb{F}_{11}^\times$ are cyclic groups.

6.7 Let G be a finite group and suppose that $H \subset G$, $H \neq \emptyset$, and for all $\alpha, \beta \in H$,

$$\alpha\beta \in H .$$

Prove that H is a subgroup of G. Suppose that $|G| = \infty$. Does this still hold? If not give a counterexample.

6.8 Find all generators of $\mathbb{Z}/60\mathbb{Z}$.

6.9 • Let p be prime. Suppose that a subgroup $H \subset S_p$ contains a transposition and an element of order p. Prove that $H = S_p$.

6.10 • Show that every subgroup H of $\mathbb{Z}$ is either trivial or of the form $n\mathbb{Z}$, where n is the least positive integer in H.

6.11 Suppose that H and K are subgroups of a group G. Show that $H \cup K$ is a subgroup if and only if $H \subset K$ or $K \subset H$.

6.12 Let G be a group and let

$$H = \{(\alpha, \alpha) \in G \times G \mid \alpha \in G\} .$$

Verify that H is a subgroup of $G \times G$ and that $H \cong G$.

6.13 Can you find two matrices that generate $SL(2, \mathbb{F}_3)$? $SL(2, \mathbb{F}_5)$? (You may use that $|SL(2, \mathbb{F}_p)| = (p-1)p(p+1)$, cf. Example 10.4(ii)).

6.14 Prove that $O(3)$ is isomorphic to $SO(3) \times \{\pm 1\}$.

6.15 • Let G be a subgroup of $O(3)$. Suppose that $-I \in G$. Prove that G is isomorphic to $G^+ \times \{\pm 1\}$, where $G^+ = G \cap SO(3)$.

6.16 Show that $\det : GL(n, F) \to F^\times$ is surjective.

6.17 Let

$$\text{sgn} : S_n \to \{\pm 1\}$$

be given by the sign of a permutation. Show that sgn is a homomorphism. Is it surjective? What is its kernel?

6.18 Let the homomorphism

$$f : (\mathbb{Z}/27\mathbb{Z})^\times \to (\mathbb{Z}/3\mathbb{Z})^\times$$

be reduction modulo 3. Write down the elements of the kernel of f. Show that it is cyclic.

6.19 • Let F be a field and let $f : F^\times \to F^\times$ be the map

$$f : a \mapsto a^2 ,$$

for $a \in F^\times$. Verify that f is a homomorphism and determine its kernel.

6.20 Show that in B_n,

(a) for $j \neq i \pm 1$, $\quad \zeta_i \zeta_j = \zeta_j \zeta_i$;

(b) for $1 < i < n$, $\quad \zeta_i \zeta_{i+1} \zeta_i = \zeta_{i+1} \zeta_i \zeta_{i+1}$.

6.21 Prove that in B_n,

(a) $\zeta_{i+1} = \rho \zeta_i \rho^{-1}$,

(b) $\zeta_{i+1} = \rho^i \zeta_1 \rho^{-i}$,

and that therefore $\{\zeta_1, \rho\}$ generates B_n.

6.22 Show that the image of S_n under the homomorphism p in Example 5.3 lies in $O(n)$, and that the image of A_n lies in $SO(n)$. Show that the image of a transposition is a reflection.

Chapter 7

Symmetry Groups

Intuitively we know when an object has symmetry and when not. Symmetry is closely related to our sense of what is beautiful. A rose window in a cathedral with its rotational symmetry is beautiful, as is a face with strong bilateral symmetry (see [5]). The octagon on the left below has no symmetry, while the one on the right is highly symmetric.

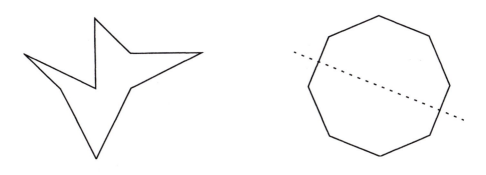

We would like a mathematically precise definition of symmetry that fits with our intuitive sense. The regular octagon above can be reflected in the dotted line. We can also rotate it about its centre through an angle of $\pi/4$. The rotation and reflection are both isometries of the Euclidean plane. This suggests a general definition.

Suppose X is an object in Euclidean n-space. A symmetry of X is an isometry that maps X to itself. We denote by $Sym(X)$ the set of all symmetries of X. Not surprisingly, $Sym(X)$ is a subgroup of $Iso(n)$. The identity map is always a symmetry of X so that $Sym(X)$ is not empty. The inverse of a symmetry is again a symmetry and so is the product of two symmetries. To say that X is highly symmetric is to say that $Sym(X)$ is large. If X has no symmetry at all, then $Sym(X)$ is the trivial group. As references for Euclidean geometry, [6], [4], and [7] are recommended.

Symmetries of Regular Polygons

Let us begin by looking at the symmetry groups of regular polygons in the plane. We will place the centre of the polygon at the origin, so that the symmetries are all elements of the orthogonal group (see [4, p. 11]).

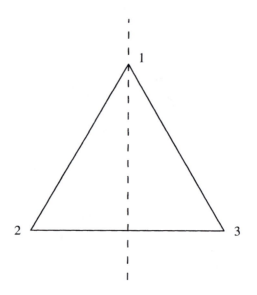

First we consider an equilateral triangle P_3. We can rotate P_3 about its centre through $2\pi/3$, $4\pi/3$, and 2π. (All angles are measured counterclockwise.) And we can reflect P_3 in the line joining a vertex to the midpoint of the opposite side. Thus $Sym(P_3)$ has six elements, three rotations, and three reflections. Notice that if we label the vertices 1, 2, 3 (counterclockwise), then the symmetries permute the 3 vertices. The rotations correspond to (1 2 3), (1 3 2), and (1), respectively. The reflections leave a vertex fixed and switch the other two. So they each correspond to a transposition. This mapping

$$Sym(P_3) \rightarrow S_3$$

is in fact an isomorphism of groups since it maps the product of two symmetries to the product of the corresponding permutations.

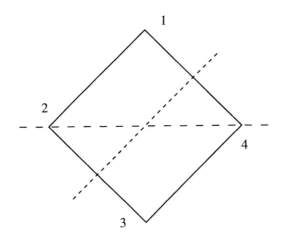

Next look at a square P_4 (see introduction of Chapter 3). We can rotate P_4 about its centre through $\pi/2$, π, $3\pi/2$, and 2π. There are two types of reflections: reflections in the diagonals, and reflections in the lines joining the midpoints of opposite sides. There are two of each type. So in all, $Sym(P_4)$ consists of four rotations and four reflections. Again we can label the vertices of P_4: 1, 2, 3, 4. Each symmetry gives a permutation of the vertices and the mapping

$$Sym(P_4) \rightarrow S_4 ,$$

defined this way, is a homomorphism: the product of two symmetries is mapped to the product of the corresponding permutations. The rotations correspond to (1 2 3 4), (1 3)(2 4), (1 4 3 2), and (1), respectively. Reflections about a diagonal correspond to the transpositions (1 3) and (2 4). The other two reflections correspond to (1 2)(3 4) and (1 4)(2 3). As we saw in Chapter 3, these are just the elements of the permutation group D_4. So we have an isomorphism

$$Sym(P_4) \xrightarrow{\;\cong\;} D_4 .$$

These examples generalize. Let P_n be a regular n-gon, $n \geq 3$. We can rotate P_n about its centre through angles $2\pi/n$, $4\pi/n$, ... , $(2n-2)\pi/n$, 2π . The reflectional symmetry of P_n depends on whether n is even or odd.

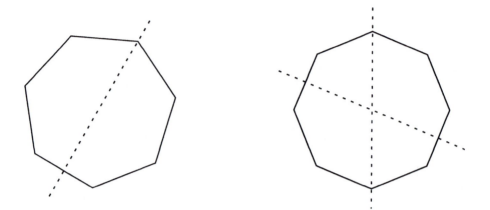

First suppose n is odd. P_n is symmetric about the line joining a vertex to the midpoint of the opposite side. There are n reflections of this type. Now suppose n is even. Then there are two types of reflections: those in a line joining the midpoints of opposite sides, and those in a diagonal joining opposite vertices. There are $n/2$ reflections of each type. In either case, $Sym(P_n)$ has n rotations and n reflections. It is called the *dihedral group* of order $2n$. We shall denote it by D_n. As in the cases $n = 3, 4$, we can label the vertices $1, 2, \ldots, n$. The symmetries of P_n permute the vertices and thus each corresponds to an element of S_n. The rotations correspond to the powers of $(1\ 2\ \cdots\ n)$. Since each symmetry is determined by its action on the vertices of P_n, this mapping from D_n into S_n is injective and gives us an isomorphism of D_n with a permutation group of degree n.

There is another way of looking at dihedral groups, independent of whether n is even or odd. Let $\sigma \in D_n$ be a rotation through $2\pi/n$ counterclockwise, and let $\tau \in D_n$ be any reflection. The n rotations are σ^j, $1 \le j \le n$. And the n reflections are $\sigma^j\tau$, $1 \le j \le n$ — this is best checked geometrically. One can also see $\sigma\tau = \tau\sigma^{-1}$. We say that σ, τ generate D_n subject to the relations

$$\sigma^n = 1 \quad \tau^2 = 1 \quad \sigma\tau = \tau\sigma^{-1} \tag{7.1}$$

because they describe multiplication in D_n completely. If n is even, then $\tau, \sigma^2\tau, \ldots$ will be the reflections of one type, and $\sigma\tau, \sigma^3\tau, \ldots$ of the other.

Symmetries of Platonic Solids

What about the symmetry groups of the Platonic solids? We shall only look at the proper symmetries (see Exercise 7.9 for the full symmetry group). This means that we first place the centre of the solid at the origin so that the isometry group will be a subgroup of $O(3)$. Then we restrict our attention to $Sym^+(X) := Sym(X) \cap SO(3)$.

Let us begin with X, the regular tetrahedron. There are two types of rotations that are symmetries of X. First, we can rotate about the line through a vertex and the centre of the opposite face. In the diagram below, the gray line is such an axis. There are two rotations, through angles $2\pi/3$ and $4\pi/3$. Since there are four vertices, there are eight such rotations in all. Secondly, we can rotate about a line like the black one, through the midpoints of opposite edges, by an angle of π. There are six edges. So this gives us three more rotations. Together with the identity, the symmetry group therefore has 12 elements. As with the regular polygons, we can regard these symmetries as permutations of the vertices of X, which we label $1, 2, 3, 4$. Then the first type of rotation corresponds to a 3-cycle, and the second one to a product of two disjoint transpositions. But these, with the identity, are just the 12 even permutations in S_4. This mapping is a homomorphism, and therefore the group of proper symmetries of the regular tetrahedron, which we denote by $\mathbb{T}$, is isomorphic to A_4.

Next we look at the proper symmetries of the cube. This is the same as the proper symmetry group of the regular octahedron, because it is just the dual of the cube. First there are the rotations about an axis like the black one below, through the centres of a pair of opposite faces, through angles $\pi/4$, $\pi/2$, and $3\pi/4$. Since there are three such pairs of faces, this gives nine rotations in all. Secondly, we can rotate the cube through angles $2\pi/3$ and $4\pi/3$ about a diagonal, like the light gray line, joining a pair of opposite vertices. There are eight vertices and therefore eight of these rotations. Thirdly, we can rotate through an angle of π about an axis like the dark gray line, through the midpoints of a pair of antipodal edges. The cube has 12 edges and thus six of these rotations. Together with the identity, this gives us 24 proper symmetries. We denote this group by $\mathbb{O}$.

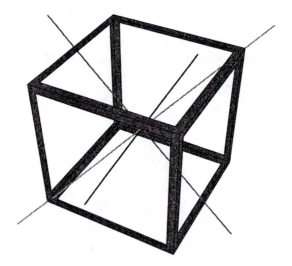

If we simply regard these symmetries as permutations of the eight vertices, we get a homomorphism of $\mathbb{O}$ into S_8, which is certainly injective. Now, $|S_8| = 40320$. So the image is a relatively small subgroup, and this homomorphism does not tell us much about $\mathbb{O}$. However, there is a more enlightening way of identifying $\mathbb{O}$ with a permutation group. Instead of the eight vertices, take the four diagonals shown below as the objects being permuted. Let us prove that this mapping into S_4 is injective. Since $|S_4| = 24$, our homomorphism is then again an isomorphism. Suppose that a proper symmetry fixes all four diagonals. Rotations of the first type do not fix any diagonals. Those of the second type fix only the diagonal that is their axis and no other. The third type of symmetry fixes no diagonal. So one that fixes all 4 must be the identity. Thus the kernel of our homomorphism is trivial and therefore it is injective.

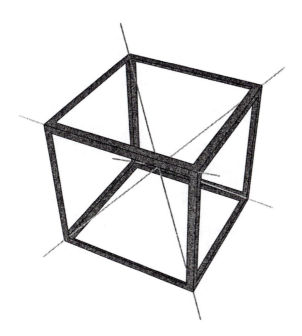

Finally, we consider the proper symmetries of the regular dodecahedron or its dual, the regular icosahedron. The regular dodecahedron has 12 faces, 30 edges, and 20 vertices. The faces are regular pentagons and three edges meet at each vertex. There are three types of rotational symmetries. First, we can rotate through angles of $2\pi/3$ and $4\pi/3$ about an axis passing through a pair of opposite vertices, like the one in light gray. There are 20 such rotations. Secondly, we can rotate through an angle of π about a line like the dark gray one passing through the midpoints of a pair of antipodal edges. We have 15 of these rotations. And lastly, we can rotate about a line like the black one through the centres of a pair of opposite faces, through angles $2\pi/5$, $4\pi/5$, $6\pi/5$, and $8\pi/5$. This gives 24 more rotations. In all we have 60 proper symmetries. Let $\mathbb{I}$ denote the proper symmetry group of the regular dodecahedron or icosahedron.

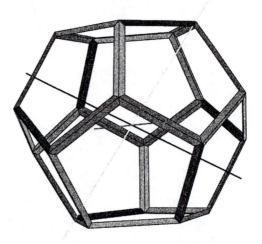

Again we can realize these symmetries as permutations. What should we take as the objects permuted? Well, the 30 edges can be grouped into 15 pairs of antipodal edges. The 15 lines through the midpoints of these pairs form five sets of mutually orthogonal triples or trihedra. In the picture below, the edges belonging to each trihedron all have the same colour. One trihedron is shown. These five trihedra are permuted by the symmetries of the icosahedron. Thus we have a homomorphism of the proper symmetry group into S_5. The rotations of order three correspond to 3-cycles. Those of order 2, to products of disjoint transpositions. Those of order five, to 5-cycles. Notice that all these permutations are even, and that in fact all 60 elements of A_5 are realized in this way. Thus $\mathbb{I}$ is isomorphic to A_5. At the publisher's web site, you will find all the drawings of the polyhedra in colour.

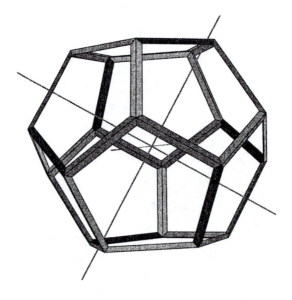

Improper Symmetries

We have computed the groups of proper symmetries of the Platonic solids. But they obviously have reflectional symmetry as well. In general, an element $\alpha \in Sym(X)$ with $\det \alpha = -1$ is called an improper symmetry. Examples are reflections in a plane and the antipodal map $-I \in O(3)$. It is easy to see that $-I \in Sym(X)$ for all five Platonic solids. Therefore by Exercise 6.15,

$$Sym(X) \cong Sym^+(X) \times \{\pm I\} \, .$$

The improper symmetries are the set $-I \cdot Sym^+(X) = -Sym(X)$.

Let us look at them in the case of the cube. Since $|\mathbb{O}| = 24$, there are also 24 improper symmetries. First of all, there are reflections. The mirrors are planes bisecting pairs of opposite faces.

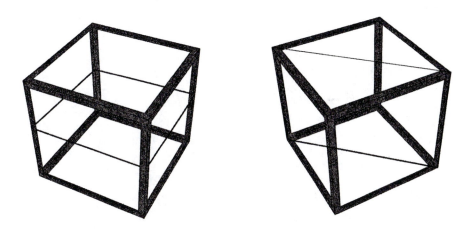

As the pictures show, there are two types. The first type gives three reflections, and the second six reflections, for a total of nine. This leaves 15 improper symmetries. They are *rotatory reflections*. A rotatory reflection is a rotation followed by a reflection in the plane perpendicular to the axis of the rotation (see [4, p. 16] or [6, §7.4]).

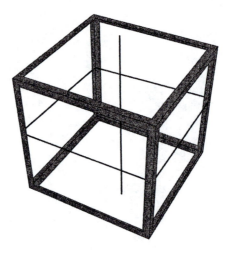

An example is a rotation through $\pi/2$ about the vertical line followed by reflection in the horizontal plane. The antipodal map is just a rotation through π about the same line followed by the reflection.

Symmetries of Equations

Symmetry also plays an important role in the study of algebraic equations. Suppose we have a polynomial equation $p(x) = 0$ with rational coefficients:

$$p(x) = x^n + a_1 x^{n-1} + \cdots + a_{n-1} x + a_n = 0$$

where $a_1, \ldots, a_n \in \mathbb{Q}$. We assume that p is irreducible, i.e., does not factor into polynomials of lower degree with rational coefficients. In the complex numbers, $p = 0$ has n roots, $z_1, \ldots, z_n$, which we will assume are distinct. We are interested in permutations of these roots that preserve any algebraic relations among them. These form a permutation group of degree n that reflects the symmetries among the roots. For a generic polynomial p there will only be "trivial" relations and the group will simply be S_n. However, for special polynomials there will be nontrivial relations and the symmetry group will not be the full permutation group.

Example 7.1
Let us look at the equation

$$0 = x^4 + x^3 + x^2 + x + 1 = \left(x^5 - 1\right)/(x - 1) .$$

The roots are

$$z_1 = e^{2\pi i/5}, \quad z_2 = e^{4\pi i/5}, \quad z_3 = e^{6\pi i/5} \quad z_4 = e^{8\pi i/5}.$$

In the picture below, they are shown on the unit circle in $\mathbb{C}$.

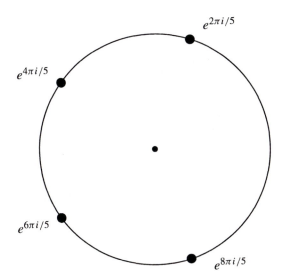

We have

$$z_2 = z_1^2, \quad z_3 = z_1^3, \quad z_4 = z_1^4.$$

We are looking for permutations of the 4 roots that preserve these relations. If α is such a permutation, then $\alpha(z_2)$, $\alpha(z_3)$, and $\alpha(z_4)$ are determined by $\alpha(z_1)$, which can be z_1, z_2, z_3, or z_4.

(i) $\alpha(z_1) = z_1$: then

$$\alpha(z_2) = \alpha(z_1)^2 = z_2 \quad \alpha(z_3) = \alpha(z_1)^3 = z_3 \quad \alpha(z_4) = \alpha(z_1)^4 = z_4.$$

So $\alpha = (1)$.

(ii) $\alpha(z_1) = z_2$: then

$$\alpha(z_2) = \alpha(z_1)^2 = z_2^2 = z_4 \quad \alpha(z_3) = z_2^3 = z_1 \quad \alpha(z_4) = z_2^4 = z_3.$$

Therefore $\alpha = (1\,2\,4\,3)$.

(iii) $\alpha(z_1) = z_3$: then

$$\alpha(z_2) = z_3^2 = z_1 \quad \alpha(z_3) = z_3^3 = z_4 \quad \alpha(z_4) = z_3^4 = z_2 \ .$$

Thus $\alpha = (1\,3\,4\,2)$.

(iv) $\alpha(z_1) = z_4$: then

$$\alpha(z_2) = z_3 \quad \alpha(z_3) = z_2 \quad \alpha(z_4) = z_1 \ .$$

So $\alpha = (1\,4)(2\,3)$.

Therefore the symmetry group of the equation is the cyclic group $\langle(1\,2\,4\,3)\rangle$. $\quad\Box$

Example 7.2
Consider the equation
$$x^4 - 10x^2 + 1 = 0 \ .$$

Its roots are
$$z_1 = \sqrt{2} + \sqrt{3}\,, \quad z_2 = -\sqrt{2} + \sqrt{3}\,, \quad z_3 = \sqrt{2} - \sqrt{3}\,, \quad z_4 = -\sqrt{2} - \sqrt{3}\,.$$

They satisfy the relations

$$\begin{array}{ll} z_1 + z_4 = 0 & \text{(i)} \\ z_2 + z_3 = 0 & \text{(ii)} \\ (z_1 + z_2)^2 = 12 & \text{(iii)} \\ (z_1 + z_3)^2 = 8 & \text{(iv)} \\ (z_2 + z_4)^2 = 8 & \text{(v)} \\ (z_3 + z_4)^2 = 12 \ . & \text{(vi)} \end{array}$$

We are interested in permutations of z_1, z_2, z_3, z_4 that preserve these relations. Well, $(1\,2)(3\,4)$, $(1\,3)(2\,4)$, $(1\,4)(2\,3)$ and (1) do. No other permutations do. To see this we check one transposition, one 3-cycle, and one 4-cycle. The others are similar. Now $(1\,2)$ takes relation (i) to

$$z_2 + z_4 = 0 \ .$$

But this contradicts relation (v). The 3-cycle $(1\,2\,3)$ does the same. The 4-cycle $(1\,2\,3\,4)$ takes (i) to

$$z_2 + z_1 = 0 \,,$$

contradicting relation (iii). Thus the symmetry group of this polynomial is the group

$$V = \{(1\,2)(3\,4),\ (1\,3)(2\,4),\ (1\,4)(2\,3),\ (1)\} \ . \quad\Box$$

The symmetry group of a polynomial equation is usually called its *Galois group*. We shall study Galois groups more extensively in later chapters.

Exercises

7.1 Look at some flowers and decide what their symmetry groups are.

7.2 What is the symmetry group of a rectangle that is not a square?

7.3 List the permutations in S_5, respectively S_6, corresponding to the elements of D_5, respectively D_6.

7.4 Let $\alpha \in D_n$ be a rotation through $2\pi/n$ counterclockwise, and let $\beta \in D_n$ be any reflection.

  (a) Prove that the n reflections in D_n are $\alpha^j \beta$, $1 \le j \le n$.
  (b) Show that $\alpha\beta = \beta\alpha^{-1}$.
  (c) Show that $\alpha^j \beta = \beta\alpha^{-j}$.

7.5 Under the given isomorphism between the group of proper symmetries of the cube and S_4, to which types of permutations do the three types of rotations correspond?

7.6 Verify that

  (a) elements of A_5 are either 5-cycles, 3-cycles, products of two disjoint transpositions, or the identity;
  (b) there are 20 3-cycles, 24 5-cycles, and 15 products of two disjoint transpositions.

7.7 In the picture below, a regular tetrahedron is inscribed in a cube. Which proper symmetries of the cube map the tetrahedron to itself?

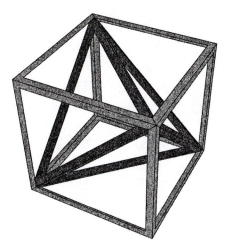

7.8 The figure below shows a cube inscribed in a regular dodecahedron. Which proper symmetries of the dodecahedron map the cube to itself?

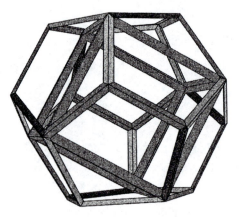

7.9 Describe the reflections that are symmetries of a regular tetrahedron and of a regular dodecahedron.

7.10 Describe the 15 rotatory reflections in the symmetry group of the cube.

7.11 Show that the symmetry group of $x^4 + 1 = 0$ is V. Suggestion: express the roots as in Example 7.1.

7.12 • Show that the symmetry group of $x^4 - 2x^2 - 2 = 0$ is D_4.

Chapter 8

Group Actions

Examples

In the previous chapter, we looked at how the symmetry group of a regular polygon permutes its vertices, and how the symmetry group of a cube permutes the four diagonals. Realizing these groups as permutation groups of a set of objects told us a lot about them. In this chapter, we are going to pursue this point of view. In general, one says that a group G acts on a set X (usually finite) if one is given a homomorphism $G \to S_X$. A more convenient way to express this is the following.

DEFINITION 8.1   *We say that a group G acts on a set X if we have a mapping $G \times X \to X$, whereby $(\alpha, x) \mapsto \alpha \cdot x \in X$, such that*

$$(\alpha\beta) \cdot x = \alpha \cdot (\beta \cdot x)$$
$$1 \cdot x = x$$

for all $\alpha, \beta \in G$ and $x \in X$.

Where there is no possibility of confusion, we shall just write αx instead of $\alpha \cdot x$.

For example, if $G = \mathbb{O}$ and $X = \{d_1, d_2, d_3, d_4\}$, the set of four diagonals of a cube, then the mapping $\mathbb{O} \times X \to X$ is given by

$$(\alpha, d_i) \mapsto \alpha(d_i) \ ,$$

where $\alpha \in \mathbb{O}$ and $1 \le i \le 4$. Here are two further examples that will interest us greatly.

Examples 8.1

(i) A group G acts on itself by multiplication on the left: for $\xi \in G$ and $\alpha \in G$, we set

$$\alpha \cdot \xi := \alpha\xi \ .$$

Then (i) is satisfied because the group operation is associative: for $\beta \in G$,

$$(\alpha\beta) \cdot \xi = (\alpha\beta)\xi = \alpha(\beta\xi) = \alpha \cdot (\beta \cdot \xi)$$

and (ii) because 1 is the identity for the group operation:

$$1 \cdot \xi = 1\xi = \xi \,.$$

(ii) A group G acts on itself by *conjugation:* for $\xi \in G$ and $\alpha \in G$, we set

$$\alpha \cdot \xi := \alpha\xi\alpha^{-1} \,.$$

Then for $\beta \in G$,

$$(\alpha\beta) \cdot \xi = (\alpha\beta)\xi(\alpha\beta)^{-1} = \alpha(\beta\xi\beta^{-1})\alpha^{-1} = \alpha \cdot (\beta \cdot \xi)$$

and

$$1 \cdot \xi = 1\xi 1^{-1} = \xi \,. \quad \square$$

It is not hard to see that this definition does give you a homomorphism $G \to S_X$. For each $\alpha \in G$, we can define a permutation σ_α of X:

$$\sigma_\alpha(x) := \alpha \cdot x \,.$$

This map σ_α is bijective because it has an inverse, namely $\sigma_{\alpha^{-1}}$:

$$\sigma_{\alpha^{-1}}\sigma_\alpha(x) = \alpha^{-1}(\alpha x) = \left(\alpha^{-1}\alpha\right) x = x$$

and therefore $\sigma_{\alpha^{-1}}\sigma_\alpha = 1$. Similarly, $\sigma_\alpha\sigma_{\alpha^{-1}} = 1$. Furthermore, the mapping $\sigma : G \to S_X$, given by

$$\sigma : \alpha \mapsto \sigma_\alpha$$

is a homomorphism. For

$$\sigma_{\alpha\beta}(x) = (\alpha\beta) \cdot x = \alpha(\beta x) = \alpha \cdot \sigma_\beta(x) = \sigma_\alpha\big(\sigma_\beta(x)\big) \,.$$

and hence $\sigma_{\alpha\beta} = \sigma_\alpha\sigma_\beta$.

Orbits and Stabilizers

If the group G acts on the set X, the *orbit* of a point $x \in X$ is the set

$$O_x := \{\alpha x \mid \alpha \in G\} \subset X \,.$$

The *stabilizer* of x is the subgroup

$$G_x := \{\alpha \in G \mid \alpha x = x\} .$$

Example 8.2
  If we take

$$X = \{1, 2, 3, 4\}$$

and

$$G = V' = \{(1), (12), (34), (12)(34)\} \subset S_4 ,$$

then

$$O_1 = O_2 = \{1, 2\}, \quad O_3 = O_4 = \{3, 4\}$$

and

$$V_1' = V_2' = \langle(34)\rangle, \quad V_3' = V_4' = \langle(12)\rangle .$$

The group

$$V = \{(1), (12)(34), (13)(24), (14)(23)\}$$

also acts on this set. However

$$O_1 = O_2 = O_3 = O_4 = X, \quad V_1 = V_2 = V_3 = V_4 = \{(1)\} . \quad \square$$

In Chapter 3 we looked at the decomposition of a permutation $\alpha \in S_n$ into a product of cycles. The cycles we found correspond to the orbits of the permutation group $\langle\alpha\rangle$ acting on the set $\{1, 2, \ldots, n\}$. The orbit of 1 is

$$O_1 = \left\{1, \alpha(1), \alpha^2(1), \ldots, \alpha^{r_1-1}(1)\right\} ,$$

which gives us the first cycle. We then pick the smallest number i_2, which does not occur in O_1, and compute its orbit:

$$O_{i_2} = \left\{i_2, \alpha(i_2), \alpha^2(i_2), \ldots, \alpha^{r_2-1}(i_2)\right\} .$$

This gives us the second cycle and so on.

In general, if a group G acting on a set X has only one orbit, we say that G acts *transitively* on X. If G is a permutation group of degree n, which acts transitively on $\{1, 2, \ldots, n\}$, we say that G is transitive.

REMARK 8.1   In general, $O_x = O_y$, for $x, y \in X$, if and only if there exists an $\alpha \in G$ such that $\alpha x = y$. Why is this so? Suppose that $\alpha x = y$. If $z = \beta y \in O_y$, then $\beta\alpha x = \beta y = z$ so that $z \in O_x$. Thus

$$O_y \subset O_x .$$

By writing $\alpha^{-1}y = x$, we can reverse the roles of x and y and see that

$$O_x \subset O_y .$$

Therefore $O_x = O_y$. Conversely, if $O_x = O_y$, then $y \in O_x$, and therefore there exists $\alpha \in G$ such that $\alpha x = y$. If there is only one orbit, then for all x and y, we have $O_x = O_y$ or equivalently, for all x and y there exists $\alpha \in G$ such that $\alpha x = y$.

∎

Many of the permutation groups that we have seen act transitively on the set $\{1, 2, \ldots, n\}$. However the stabilizers may not be trivial. For example, taking $G = S_n$, we see that any permutation that fixes 1 can permute $\{2, \ldots, n\}$ quite arbitrarily. So $(S_n)_1$ is the group of all permutations of $\{2, \ldots, n\}$, which is isomorphic to S_{n-1}.

The group of proper symmetries of a Platonic solid acts transitively on the set of vertices, the set of edges, and the set of faces. What do the stabilizers look like? Let us look at the regular tetrahedron. First pick a vertex. The rotations, whose axes pass through the vertex and the centre of the opposite face, leave it fixed. No other rotations of order 3 do so. A rotation of order 2, whose axis passes through the centres of a pair of opposite edges, does not fix any vertex. So the stabilizer of a vertex is the cyclic group of order 3, consisting of the rotations about the line through the vertex and the centre of the opposite side. The only rotation that fixes a given edge is the half-turn about the line joining its midpoint and that of the opposite edge. So its stabilizer is the cyclic group of order 2 generated by this half-turn. And the stabilizer of a face is just the stabilizer of the opposite vertex.

In Example 8.1(i) above, G acts transitively on itself. Why? Take $v, \xi \in G$. Let $\alpha = \xi v^{-1}$. Then

$$\alpha v = \left(\xi v^{-1} \right) v = \xi \ .$$

The stabilizer of any element ξ is trivial because $\alpha \xi = \xi$ implies that $\alpha = 1$.

The action of G on itself by conjugation (Example 8.1(ii)) is more interesting: first some terminology. The orbit of an element $\xi \in G$ under conjugation is called its *conjugacy class* and will be denoted by C_ξ. So

$$C_\xi := \{\alpha \xi \alpha^{-1} \mid \alpha \in G\} \ .$$

An element $\alpha \xi \alpha^{-1}$ is called a conjugate of ξ. The stabilizer of ξ is called its *centralizer*, denoted by $Z_\xi(G)$ or just Z_ξ:

$$Z_\xi := \{\alpha \in G \mid \alpha \xi \alpha^{-1} = 1\} = \{\alpha \mid \alpha \xi = \xi \alpha\} \ .$$

Thus the centralizer of ξ is the set of all elements that commute with ξ. The *centre* of G, written $Z(G)$, is

$$Z(G) = \{\alpha \mid \alpha \beta = \beta \alpha, \text{ for all } \beta \in G\} \ .$$

So

$$Z(G) = \cap_\xi Z_\xi(G) \ .$$

If G is abelian, then of course $Z(G) = G$.

Let us compute the conjugacy classes of S_n.

LEMMA 8.1

Suppose $v = (i_1, \ldots, i_r) \in S_n$ is an r-cycle. Then $\alpha v \alpha^{-1}$ is the r-cycle $(\alpha(i_1), \ldots, \alpha(i_r))$.

PROOF   We have

$$\alpha(i_j) \xrightarrow{\alpha^{-1}} i_j \xrightarrow{v} i_{j+1} \xrightarrow{\alpha} \alpha(i_{j+1}) \, ,$$

where as usual, $i_{r+1} := i_1$. (Remember that we are reading from right to left.) If $i \notin \{\alpha(i_1), \ldots, \alpha(i_r)\}$, then $\alpha^{-1}(i) \notin \{i_1, \ldots, i_r\}$, so that

$$i \xrightarrow{\alpha^{-1}} \alpha^{-1}(i) \xrightarrow{v} \alpha^{-1}(i) \xrightarrow{\alpha} i \, .$$

Thus

$$\alpha v \alpha^{-1} = (\alpha(i_1), \ldots, \alpha(i_r)) \, . \quad \blacksquare$$

It also follows from this lemma that if we are given a second r-cycle ξ, there exists an $\alpha \in G$ such that $\xi = \alpha v \alpha^{-1}$. Namely, if $\xi = (j_1, \ldots, j_r)$, then define

$$\alpha(i_1) = j_1, \ldots, \alpha(i_r) = j_r \, ,$$

and extend α to the rest of $\{1, 2, \ldots, n\}$ in any way you like as long as α is bijective. Then by the lemma, $\xi = \alpha v \alpha^{-1}$. Thus the set of all r-cycles is a conjugacy class.

Now suppose that $v = v_1 \cdots v_s$ where $v_1, \ldots, v_s$ are disjoint cycles of length $r_1, \ldots, r_s$, respectively. We say that v is of *cycle type* $\{r_1, \ldots, r_s\}$. Then for any $\alpha \in G$, we have

$$\xi := \alpha v \alpha^{-1} = \alpha v_1 \alpha^{-1} \cdots \alpha v_s \alpha^{-1} \, .$$

Therefore ξ too is of cycle type $\{r_1, \ldots, r_s\}$. On the other hand, given any two elements v and ξ of the same cycle type, we can refine the argument just given to see that there exists an α such that $\xi = \alpha v \alpha^{-1}$. This proves the following theorem:

THEOREM 8.1

A conjugacy class in S_n consists of all permutations of a given cycle type.

For example, the possible cycle types in S_5 are

$$\{1\} \quad \{2\} \quad \{3\} \quad \{2, 2\} \quad \{4\} \quad \{2, 3\} \quad \{5\} \, .$$

To each of these there corresponds a conjugacy class in S_5. Notice that a cycle type is given by any set $\{r_1, \ldots, r_s\} \subset \mathbb{N}$, where $r_1 + \cdots + r_s = n$. Such a set is called a *partition* of n. The list above can also be regarded as the possible partitions of 5. As with cycle decompositions, we write $\{3\}$ instead of $\{3, 1, 1\}$ and so on.

Fractional Linear Transformations

In this section we will look at an action of $GL(2, F)$ on the projective line. More will be said about it in Example 10.4(iii) and it will be important in Chapter 12. You may well be familiar with *fractional linear transformations* over the complex numbers. Given a matrix

$$\alpha = \begin{pmatrix} a & b \\ c & d \end{pmatrix} \in GL(2, \mathbb{C}) ,$$

one defines

$$s_\alpha(z) = \frac{az + b}{cz + d} ,$$

which takes on complex values for $z \in \mathbb{C}$, except at $z = -d/c$. To deal with this value, one extends s_α to the extended complex line (or Riemann sphere), $P(\mathbb{C}) = \mathbb{C} \cup \{\infty\}$, by setting

$$s_\alpha(-d/c) = \infty, \quad s_\alpha(\infty) = a/c .$$

The reason for the latter formula is, that if you write

$$s_\alpha(z) = \frac{a + b/z}{c + d/z} ,$$

and then set $1/z = 0$, you obtain a/c.

This works perfectly well for any field F, not just $\mathbb{C}$. The extended affine line, or projective line, over F can be defined formally by

$$P(F) := F \cup \{\infty\} .$$

Then for

$$\alpha = \begin{pmatrix} a & b \\ c & d \end{pmatrix} \in GL(2, F) ,$$

we set

$$s_\alpha(x) := \frac{ax + b}{cx + d} , \qquad x \in F , \ x \neq -d/c ,$$

and

$$s_\alpha(-d/c) = \infty, \quad s_\alpha(\infty) = a/c .$$

It is easy to check that we have defined an action of $GL(2, F)$ on $P(F)$. This action is transitive because we can see that $O_0 = P(F)$: take any $b \in F$, then

$$s_\alpha(0) = b \quad \text{for} \quad \alpha = \begin{pmatrix} 1 & b \\ 0 & 1 \end{pmatrix}$$

and

$$s_\alpha(0) = \infty \quad \text{for} \quad \alpha = \begin{pmatrix} 0 & 1 \\ 1 & 0 \end{pmatrix} .$$

The action is particularly interesting when $F = \mathbb{F}_p$ for some prime p. We can write $P(\mathbb{F}_p)$ as

$$P(\mathbb{F}_p) = \{0, 1, \ldots, p - 1, \infty\}$$

and can regard each s_α as a permutation of this set of $p + 1$ elements. In this way the action gives us a homomorphism $f_p : GL(2, \mathbb{F}_p) \to S_{p+1}$, with $f_p(\alpha) := s_\alpha$.

THEOREM 8.2

$$\ker f_p = \mathbb{F}_p^\times \cdot I = Z\big(GL(2, \mathbb{F}_p)\big)$$

PROOF   Clearly, $\mathbb{F}_p^\times \cdot I \subset \ker f_p$. Suppose that $\alpha \in \ker(f_p)$. Then $s_\alpha(x) = x$ for all $x \in P(\mathbb{F}_p)$. In particular this holds for $x = 0, 1, \infty$. This means that α lies in the stabilizers of $0, 1$, and ∞. Let us compute the stabilizer of 0: we have

$$0 = s_\alpha(0) = \frac{a \cdot 0 + b}{c \cdot 0 + d} = \frac{b}{d}$$

if and only if $b = 0$. So the stabilizer of 0 is

$$\left\{ \begin{pmatrix} a & 0 \\ c & d \end{pmatrix} \in GL(2, \mathbb{F}_p) \right\} .$$

Similarly,

$$\infty = s_\alpha(\infty) = \frac{a}{c}$$

if and only if $c = 0$. Thus the stabilizer of ∞ is

$$\left\{ \begin{pmatrix} a & b \\ 0 & d \end{pmatrix} \in GL(2, \mathbb{F}_p) \right\} .$$

So if α fixes both 0 and ∞, then

$$\alpha = \begin{pmatrix} a & 0 \\ 0 & d \end{pmatrix} ,$$

for some $a, d \in \mathbb{F}_p^\times$. If, in addition, α fixes 1, then

$$1 = \alpha(1) = a/d ,$$

and therefore $a = d$. Thus $\alpha = aI$ for some $a \in \mathbb{F}_p$, and

$$\ker(f_p) = \left\{ aI \mid a \in \mathbb{F}_p^\times \right\} .$$

In Exercise 8.11 it is shown that this is the centre of $GL(2, \mathbb{F}_p)$.   ∎

Let us look at this map for $p = 2$. If we take

$$\alpha = \begin{pmatrix} 1 & 1 \\ 0 & 1 \end{pmatrix}, \quad \text{we have} \quad s_\alpha(x) = x + 1 \, .$$

Thus

$$s_\alpha(0) = 1, \quad s_\alpha(1) = 0, \quad s_\alpha(\infty) = \infty \, .$$

So

$$f_2(\alpha) = (0 \ 1) \, .$$

For

$$\beta = \begin{pmatrix} 0 & 1 \\ 1 & 0 \end{pmatrix}, \quad \text{we have} \quad s_\beta(x) = \frac{1}{x} \, .$$

Thus

$$s_\beta(0) = \infty, \quad s_\beta(1) = 1, \quad s_\beta(\infty) = 0 \, .$$

So

$$f_2(\beta) = (0 \ \infty) \, .$$

As we saw in Chapter 3, these two transpositions generate S_3. So the image of f_2 must be all of S_3. Since

$$\ker(f_2) = \left\{ aI \mid a \in \mathbb{F}_2^\times \right\} = \{ I \} \, ,$$

f_2 is injective as well, and therefore an isomorphism.

Now let us consider f_3. We know that

$$|GL(2, \mathbb{F}_3)| = 48 \quad \text{and} \quad |S_4| = 24 \, ,$$

and

$$\ker(f_2) = \{ I, 2I \} \, .$$

So f_3 is not an isomorphism. However, it is surjective. We can again find matrices mapping onto three transpositions that generate S_4. Take

$$\alpha = \begin{pmatrix} 2 & 1 \\ 0 & 1 \end{pmatrix} \qquad \beta = \begin{pmatrix} 2 & 2 \\ 0 & 1 \end{pmatrix} \qquad \gamma = \begin{pmatrix} 0 & 1 \\ 1 & 0 \end{pmatrix} \, .$$

Then

$$s_\alpha(x) = 2x + 1 \quad s_\beta(x) = 2x + 2 \quad s_\gamma(x) = \frac{1}{x}$$

and

$$f_3(\alpha) = (0 \ 1) \qquad f_3(\beta) = (0 \ 2) \qquad f_3(\gamma) = (0 \ \infty) \, .$$

The action of $GL(2, F)$ on $P(F)$ is more than just transitive. In fact given two triples of distinct points in $P(F)$, there exists an $\alpha \in GL(2, F)$ such that s_α maps one triple to the other. To see this we shall show that given three distinct elements, $u, v, w \in P(F)$, we can find an α such that

$$s_\alpha(0) = u \qquad s_\alpha(1) = v \qquad s_\alpha(\infty) = w .$$

Let

$$\alpha = \begin{pmatrix} a & b \\ c & d \end{pmatrix} .$$

Then the three equations above are

$$s_\alpha(0) = b/d = u \qquad s_\alpha(1) = (a+b)/(c+d) = v \qquad s_\alpha(\infty) = a/c = w .$$

Substituting the first and third into the second, we get

$$(cw + du)/(c + d) = v$$

which has the solution

$$c = d(v - u)/(w - v) ,$$

for $d \in F^\times$. Thus we have a solution α unique up to multiplication by a nonzero scalar, in other words, by a matrix in the centre of $GL(2, F)$. This means that s_α is uniquely determined.

If a group G acts on a set X and maps any distinct triple of points in X to any other, then we say that the action is triply transitive. If it maps any distinct pair of points to any other, it is called doubly transitive. So the action of $GL(2, F)$ on $P(F)$ is triply transitive and you can check that the action of $SL(2, F)$ on $P(F)$ is doubly transitive (see Exercise 8.24).

Cayley's Theorem

As we saw in the first section, defining an action of a group G on a set X, is the same as giving a homomorphism

$$\sigma : G \to S_X .$$

The kernel of σ is

$$\{\alpha \in G \mid \alpha x = x \text{ for all } x \in X\} ,$$

in other words, those elements of G that act trivially on X. It is sometimes called the *kernel* of the action.

Now suppose that G is a finite group, and let G act on itself by multiplication on the left. Then the mapping σ is a homomorphism from $G \to S_G$. And

$$\ker(\sigma) = \{\alpha \in G \mid \alpha\xi = \xi \text{ for all } \xi \in G\} .$$

But taking $\xi = 1$, we see that such an α must be 1. So σ is injective. This gives us a result known as Cayley's Theorem:

THEOREM 8.3
Let G be a finite group of order n. Then G is isomorphic to a permutation group of degree n, more precisely to a subgroup of the group of permutations of G itself.

Software and Calculations

The function Orbit [G, x] will compute the orbit of x under the permutation group G. Here x is a positive natural number or a vector. For example,

```
In[1] := A5 = Group[ P[{1,2,3}], P[{3,4,5}] ]
```

$$Out[1] = \langle (1, 2, 3), (3, 4, 5) \rangle$$

So the orbit of 2 under A_5 can be computed by

```
In[2] := Orbit[A5, 2]
```

$$Out[2] = \{2, \ 3, \ 1, \ 4, \ 5\}$$

Similarly, if you set

```
In[3] := ChoosePrime[5]
```

$$Out[3] = 5$$

then

```
In[4] := F20 = Group[ L[{1,1},{0,1}], L[{2,0},{0,3}] ]
```

$$Out[4] = \left\langle \begin{pmatrix} 1 & 1 \\ 0 & 1 \end{pmatrix}, \begin{pmatrix} 2 & 0 \\ 0 & 3 \end{pmatrix} \right\rangle$$

and the orbit of the vector (2, 3) is

```
In[5] := Orbit[F20, {2,3}]
```

$$Out[5] = \left\{ \begin{pmatrix} 2 \\ 3 \end{pmatrix}, \begin{pmatrix} 0 \\ 3 \end{pmatrix}, \begin{pmatrix} 0 \\ 4 \end{pmatrix}, \begin{pmatrix} 0 \\ 2 \end{pmatrix}, \begin{pmatrix} 0 \\ 1 \end{pmatrix}, \begin{pmatrix} 1 \\ 1 \end{pmatrix}, \begin{pmatrix} 2 \\ 1 \end{pmatrix}, \begin{pmatrix} 2 \\ 2 \end{pmatrix}, \right.$$

$$\begin{pmatrix} 3 \\ 1 \end{pmatrix}, \begin{pmatrix} 1 \\ 3 \end{pmatrix}, \begin{pmatrix} 2 \\ 4 \end{pmatrix}, \begin{pmatrix} 1 \\ 4 \end{pmatrix}, \begin{pmatrix} 3 \\ 3 \end{pmatrix}, \begin{pmatrix} 4 \\ 1 \end{pmatrix}, \begin{pmatrix} 4 \\ 2 \end{pmatrix}, \begin{pmatrix} 1 \\ 2 \end{pmatrix},$$

$$\left. \begin{pmatrix} 3 \\ 2 \end{pmatrix}, \begin{pmatrix} 4 \\ 3 \end{pmatrix}, \begin{pmatrix} 3 \\ 4 \end{pmatrix}, \begin{pmatrix} 4 \\ 4 \end{pmatrix} \right\}$$

(see Exercise 4.4).

Stabilizer[G,x] will compute the stabilizer of x in the group G. So for example

```
In[6] := Stabilizer[A5, 3]

Out[6] = ⟨ (1 2 4), (1 2 5) ⟩
```

The conjugacy class of an element can be calculated with the function ConjugacyClass. Let us use it to compute the conjugacy classes in A_5. To begin with, we know that the conjugacy classes in S_5 correspond to the cycle types of permutations of degree 5. The even cycle types are

$$\{1\}, \{3\}, \{5\}, \{2, 2\} .$$

The function CycleTypes computes the number of permutations in each cycle type:

```
In[7] := CycleTypes[A5]
```

$$Out[7] = \begin{array}{cccc} \{1\} & \{3\} & \{5\} & \{2, 2\} \\ 1 & 20 & 24 & 15 \end{array}$$

Now two elements in A_5 may be conjugate by an element in S_5, but not by an element in A_5. So a conjugacy class of S_5 may break up into more than one conjugacy class in A_5. We begin with a 3 -cycle:

```
In[8] := ConjugacyClass[ A5, P[{1,2,3}] ]

Out[8] = { (1 2 3), (1 2 4), (1 2 5), (2 3 4),
           (1 4 3), (1 5 4), (1 3 5), (1 5 2),
           (1 3 2), (1 4 2), (2 3 5), (1 5 3),
           (1 3 4), (1 4 5), (2 4 3), (2 4 5),
           (2 5 3), (2 5 4), (3 4 5), (3 5 4)}
```

These are all 20 3-cycles. Next we look at the conjugacy class of a 5-cycle:

```
In[9] := ConjugacyClass[ A5, P[{1,2,3,4,5}] ]

Out[9] = {(1 2 3 4 5), (1 2 4 5 3), (1 2 5 3 4),
          (1 4 2 3 5), (1 4 5 2 3), (1 5 2 4 3),
          (1 3 2 5 4), (1 3 5 4 2), (1 4 3 5 2),
          (1 5 3 2 4), (1 3 4 2 5), (1 5 4 3 2)}
```

This is only half of the 5-cycles! One that is missing is (1 2 3 5 4) . So let us compute its conjugacy class:

```
In[10] := ConjugacyClass[ A5, P[{1,2,3,5,4}] ]

Out[10] = {(1 2 3 5 4), (1 2 4 3 5), (1 2 5 4 3),
           (1 5 2 3 4), (1 3 2 4 5), (1 3 4 5 2),
           (1 4 2 5 3), (1 4 5 3 2), (1 5 3 4 2),
           (1 4 3 2 5), (1 3 5 2 4), (1 5 4 2 3)}
```

These are the remaining 5-cycles. Lastly we look at the conjugacy class of a product of two transpositions:

```
In[11] := ConjugacyClass[ A5, P[{1,2},{3,4}] ]

Out[11] = {(1 2)(3 4), (1 2)(4 5), (1 2)(3 5),
           (1 4)(2 3), (1 3)(2 4), (1 4)(2 5),
           (1 5)(2 3), (1 3)(2 5), (1 5)(2 4),
           (2 3)(4 5), (1 3)(4 5), (1 4)(3 5),
           (1 5)(3 4), (2 4)(3 5), (2 5)(3 4)}
```

These are all 15 permutations of type $\{2, 2\}$. So these four sets together with $\{(1)\}$ are the conjugacy classes of A_5.

The centre of a group can be computed with the function Centre. For example,

```
In[12] := D4 = Group[ P[{1, 2, 3, 4}], P[{1, 3}] ]

Out[12] = ⟨(1, 2, 3, 4), (1, 3)⟩
```

And

```
In[13] := Centre[D4]

Out[13] = ⟨(1, 3)(2, 4)⟩
```

For a matrix α in $GL(2, \mathbb{F}_p)$ the corresponding fractional linear transformation s_α is computed by the function FLTPermutation . Let us repeat the calculation of $f_3 : GL(2, \mathbb{F}_3) \to S_4$ using this function:

In[14]:= ChoosePrime[3]

Out[14]= 3

In[15]:= a = L[{{2,1},{0,1}}]

Out[15]= $\begin{pmatrix} 2 & 1 \\ 0 & 1 \end{pmatrix}$

In[16]:= b = L[{{2,2},{0,1}}]

Out[16]= $\begin{pmatrix} 2 & 2 \\ 0 & 1 \end{pmatrix}$

In[17]:= c = L[{{0,1},{1,0}}]

Out[17]= $\begin{pmatrix} 0 & 1 \\ 1 & 0 \end{pmatrix}$

In[18]:= FLTPermutation[a]

Out[18]= (0, 1)

In[19]:= FLTPermutation[b]

Out[19]= (0, 2)

In[20]:= FLTPermutation[c]

Out[20]= (0, ∞)

Exercises

8.1 What is the stabilizer in D_n of the vertex of a regular n-gon?

8.2 What are the stabilizers of a vertex, an edge, and a face of a cube in the octahedral group? Of a regular dodecahedron in the icosahedral group?

8.3 Describe the conjugacy classes of S_6.

8.4 Compute the conjugacy classes of A_6.

8.5 What are the conjugacy classes of D_5?

8.6 Determine the conjugacy classes in $SL(2, \mathbb{F}_5)$.

8.7 Prove that the conjugacy class of an element $\alpha \neq 1$ in $SO(3)$ is uniquely determined by

  (a) a unit vector $v \in \mathbb{R}^3$ such that the axis of rotation of α is the line through v, and

  (b) an angle of rotation $t \in (0, \pi]$, whereby the conjugacy class corresponding to (v, π) is the same as the one corresponding to $(-v, \pi)$.

8.8 What is the centralizer of an r-cycle in S_n?

8.9 • A group G acts on a set X. Suppose that $x, y \in X$, and $y = \alpha x$ for some $\alpha \in G$. Prove that

$$G_y = \alpha G_x \alpha^{-1} := \left\{ \alpha \beta \alpha^{-1} \mid \beta \in G_x \right\} .$$

8.10 Verify that $(\alpha, x) \mapsto s_\alpha(x)$, $\alpha \in GL(2, F)$, $x \in P(F)$ defines an action of $GL(2, F)$ on $P(F)$.

8.11 Prove that the centre of $GL(2, F)$ is $\{aI \mid a \in F^\times\}$.

8.12 • Find the centre of D_n, $n \geq 3$.

8.13 Let G be a group. For any $\alpha \in G$, define

$$c_\alpha : G \to G$$

by

$$c_\alpha(\beta) = \alpha \beta \alpha^{-1} .$$

Prove that c_α is an automorphism of G (see Exercise 5.21). Thus the conjugate of a product is the product of the conjugates, and the conjugate of an inverse is the inverse of the conjugate. Such an automorphism is called an *inner automorphism*.

8.14 Define a map
$$c : G \rightarrow \mathrm{Aut}(G)$$
by
$$c(\alpha) = c_\alpha .$$
Check that c is a homomorphism. What is its kernel?

8.15 • Let
$$H = \left\{ \begin{pmatrix} 1 & a & b \\ 0 & 1 & c \\ 0 & 0 & 1 \end{pmatrix} \middle| a, b, c \in \mathbb{R} \right\} .$$
Verify that H is a linear group (H is called the *Heisenberg group*). Compute its centre.

8.16 • Let G_{72} be the permutation group generated by $g = \{(1\,2\,3),\ (1\,4)(2\,5)(3\,6),$ $(1\,5\,2\,4)(3\,6)\}$. It has order 72. Verify that G_{72} is transitive. Determine the stabilizer of 1. Show that it is isomorphic to $\mathbb{Z}/2\mathbb{Z} \times S_3$.

8.17 • Find the transitive subgroups of S_4. Suggestion: first check the subgroups generated by at most two elements.

8.18 Is f_5 surjective? Is f_p surjective for any primes $p > 5$?

8.19 • Show that f_5 is injective on F_{20} (see Exercise 4.4) and that its image lies in the stabilizer of ∞. Thus it can be identified with a permutation group of degree 5. Verify that in S_5 it can be generated by $\{(1\,2\,3\,4\,5),\ (1\,2\,4\,3)\}$.

8.20 Are two elements in F_{20}, which are conjugate in S_5, also conjugate in F_{20} itself?

8.21 Show that $SL(2, \mathbb{Z})$ acts transitively on $P(\mathbb{Q})$. Does it act doubly transitively?

8.22 • Let the Frobenius group $F_{p(p-1)}$ (see Exercise 4.4) act by fractional linear transformations on $P(\mathbb{F}_p)$.

(a) Verify that for $a \in \mathbb{F}_p^\times$ and $b \in \mathbb{F}_p$, the matrix
$$\begin{pmatrix} a & b \\ 0 & 1 \end{pmatrix}$$
acts by the mapping $f_{a,b}$:
$$f_{a,b}(x) = ax + b , \quad x \in \mathbb{F}_p .$$
Check that $F_{p(p-1)}$ fixes ∞.

(b) Show that
$$f_{a,b} \circ f_{c,d} = f_{ac,ad+b} .$$

   (c) Show that $F_{p(p-1)}$ acts transitively on $P(\mathbb{F}_p) \setminus \{\infty\}$. Does it act doubly transitively?

8.23 Suppose that G acts transitively on X. Show that G acts doubly transitively if and only if G_x acts transitively on $X \setminus \{x\}$ for some $x \in X$.

8.24 • Show that $SL(2, F)$ acts doubly transitively on $P(F)$ for any field F.

Chapter 9

Counting Formulas

The Class Equation

If a group G acts on a set X, then X breaks up into the disjoint union of the various orbits of G. When X and G are finite, we can obtain formulas relating the number of elements in the orbits and stabilizers and in X, and for the number of fixed points of the elements of G. These formulas are useful in studying the structure of abstract finite groups and of symmetry groups. They also have applications to combinatorial problems. For any finite set Y, we shall denote the number of elements in Y by $|Y|$.

Recall that in Example 8.2, we looked at the actions of V' and V on $X = \{1, 2, 3, 4\}$ and determined the orbits and stabilizers. For V' we found that

$$|V'_x| = 2 \quad \text{and} \quad |O_x| = 2,$$

and for V

$$|V_x| = 1 \quad \text{and} \quad |O_x| = 4,$$

for any $x \in X$. So in both cases $|G_x||O_x| = |G|$. This relation holds in general.

Suppose we have a group G acting on a set X. Fix a point $x \in X$, and define a map

$$e : G \to O_x$$

by

$$e(\alpha) = \alpha \cdot x,$$

for $\alpha \in G$. This map is surjective by the definition of the orbit of x. When do two elements $\alpha, \beta \in G$ have the same image y under e? Well, $\alpha x = \beta x = y$ means that $\alpha^{-1}\beta x = x$ or equivalently, that

$$\gamma := \alpha^{-1}\beta \in G_x.$$

On the other hand, if $\gamma \in G_x$, then

$$(\alpha\gamma)x = \alpha x = y.$$

Thus

$$e^{-1}(y) = \alpha \, G_x := \{\alpha\gamma \mid \gamma \in G_x\}$$

where $\alpha x = y$.

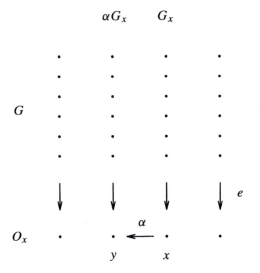

Now suppose that $|G|$ is finite. We have a bijection

$$G_x \leftrightarrow e^{-1}(y)$$

given by $\gamma \leftrightarrow \alpha\gamma$, so that

$$|e^{-1}(y)| = |G_x| \,.$$

Since

$$G = \bigsqcup_{y \in O_x} e^{-1}(y) \,,$$

it follows that

$$|G| = \sum_{y \in O_x} |e^{-1}(y)| = \sum_{y \in O_x} |G_x| = |O_x||G_x| \,.$$

This gives us our formula.

FORMULA 9.1
$|G| = |O_x||G_x| \,.$

Typically one uses this to compute $|G_x|$. For example in Exercise 8.16, the group $G = G_{72}$ acts transitively on $\{1, 2, 3, 4, 5, 6\}$. Therefore $|G_1| = 72/6 = 12$. If we let X be the set of vertices of a cube, and $G = \mathbb{O}$, then G acts transitively, so that the order of any stabilizer is $24/8 = 3$. Notice that this equation says that $|O_x|$ always divides $|G|$.

Next we look at a formula for $|X|$ in terms of data about the orbits. There is an equivalence relation hiding in any group action. Define

$$x \sim y \quad \text{if} \quad O_x = O_y .$$

This relation is reflexive, symmetric, and transitive, and therefore is an equivalence relation. By Remark 8.1,

$$x \sim y \quad \text{if and only if} \quad \alpha x = y ,$$

for some $\alpha \in G$. The equivalence class of an element x is its orbit O_x. Since distinct equivalence classes are disjoint, it follows that distinct orbits are disjoint. If X is also finite, then we can obtain a count of the elements of X. Let $x_1, \ldots, x_r$ be representatives of the orbits of G. Then

$$X = \bigsqcup_{i=1}^{r} O_{x_i} .$$

Combining this with Formula 9.1, we have

FORMULA 9.2

$$|X| = \sum_{i=1}^{r} |O_x| = \sum_{i=1}^{r} |G|/|G_{x_i}| .$$

Let us consider this formula in the case where G acts on itself by conjugation. So the orbits are the conjugacy classes. If an element $\alpha \in Z(G)$, then its conjugacy class is just $\{\alpha\}$ because α commutes with every element of G. Conversely, any element whose conjugacy class has only one element in it must commute with all elements of G and therefore lies in $Z(G)$. Let $\alpha_1, \ldots, \alpha_r$ be representatives of the other, nontrivial conjugacy classes. Then Formula 9.2 gives us the following result:

THEOREM 9.1 (The Class Equation)

$$|G| = |Z(G)| + \sum_{j=1}^{r} |C_{\alpha_j}| = |Z(G)| + \sum_{j=1}^{r} |G|/|Z_{\alpha_j}| .$$

Examples 9.1

(i) Take $G = S_4$. We know that the conjugacy classes are given by the cycle types. The possible cycle types are

$$\{1\} \quad \{2\} \quad \{3\} \quad \{4\} \quad \{2, 2\} \,.$$

The number of elements in the corresponding conjugacy classes are 1, 6, 8, 6, and 3, respectively. So the class equation is

$$24 = 1 + 6 + 8 + 6 + 3 \,.$$

(ii) In the last section of the previous chapter, we computed the conjugacy classes of A_5 directly. We now want to compute this calculation in a different way. As pointed out in the previous chapter, the possible cycle types are

$$\{1\} \quad \{3\} \quad \{5\} \quad \{2, 2\} \,.$$

These correspond to conjugacy classes in S_5. We must decide whether two permutations, which are conjugate in S_5, are also conjugate in A_5. For this we need the following observation.

REMARK 9.1   If a group G acts on a set X and $H < G$, then $H_x = G_x \cap H$ for any $x \in X$.   ∎

Recall that $|S_5| = 120$ and $|A_5| = 60$. First, we check whether the set of 3-cycles is a conjugacy class in A_5. The number of 3-cycles is 20. Therefore the centralizer of a 3-cycle, say $\xi = (1\,2\,3)$, in S_5 has order $120/20 = 6$ by Formula 9.1. Now $(4\,5)$ commutes with $(1\,2\,3)$ and so does $(1\,2\,3)$ itself. But $\langle (1\,2\,3), (4\,5) \rangle$ has order 6. So

$$Z_\xi = \langle (1\,2\,3), (4\,5) \rangle \,.$$

The intersection
$$Z_\xi \cap A_5 = \langle (1\,2\,3) \rangle \,,$$

which has order 3. Therefore the order of the conjugacy class of ξ in A_5 is $60/3 = 20$. So the set of 3-cycles is a single conjugacy class in A_5 too.

Next, let us look at the set of 5-cycles. There are 24 of them, which means that the centralizer of one of them in S_5 has order $120/24 = 5$. So this centralizer is just the cyclic subgroup generated by the 5-cycle itself. This subgroup lies in A_5. Therefore, the order of the conjugacy class of the 5-cycle in A_5 is $60/5 = 12$. So the set of 5-cycles breaks up into two conjugacy classes in A_5.

Lastly, we make the calculation for the set of all products of two disjoint 2-cycles. There are 15 of them, so the order of the centralizer of one of them in S_5

is 8. Again, we pick one, say, $\xi = (1\,2)(3\,4)$. We find that $(1\,3\,2\,4)$ commutes with it since $(1\,3\,2\,4)^2 = (1\,2)(3\,4)$. So do $(1\,3)(2\,4)$ and $(1\,4)(2\,3)$. This gives us a subgroup of order 8, which must be the centralizer of ξ in S_5:

$$Z_\xi = \langle (1\,4)(2\,3), (1\,3\,2\,4) \rangle .$$

Now

$$Z_\xi \cap A_5 = \langle (1\,4)(2\,3), (1\,2)(3\,4) \rangle \cong V .$$

Therefore the conjugacy class of $(1\,2)(3\,4)$ in A_5 has order $60/4 = 15$ as well, and is the set of all products of two disjoint transpositions. So the class equation is

$$60 = 1 + 20 + 12 + 12 + 15 .$$

(iii) Take $G = D_n$. In chapter 7 we saw that

$$D_n = \left\{ 1, \sigma, \dots, \sigma^{n-1}, \tau, \sigma\tau, \dots, \sigma^{n-1}\tau \right\} ,$$

where σ is a rotation and τ is a reflection that satisfy the relations

$$\sigma^n = 1 \quad \tau^2 = 1 \quad \sigma\tau = \tau\sigma^{-1} .$$

It follows that

$$\sigma \left(\sigma^j \tau \right) \sigma^{-1} = \sigma^{j+2}\tau ,$$

where the index j is taken modulo n, and

$$\tau \sigma^j \tau = \sigma^{n-j} .$$

Using this let us work out what the conjugacy classes are for $n = 4$ and $n = 5$. For $n = 4$, this tells us that $Z(D_4) = \{1, \sigma^2\}$ and that the other conjugacy classes are

$$\left\{ \tau, \sigma^2\tau \right\} \qquad \left\{ \sigma\tau, \sigma^3\tau \right\} \qquad \left\{ \sigma, \sigma^3 \right\} .$$

So the class equation is

$$8 = 2 + 2 + 2 + 2 .$$

For $n = 5$, the centre is trivial. The conjugacy class of τ is

$$\left\{ \tau, \sigma^2\tau, \sigma^4\tau, \sigma\tau, \sigma^3\tau \right\} .$$

Since the order of every conjugacy class must divide 10, the remaining classes must each have order 2. So they are

$$\left\{ \sigma, \sigma^4 \right\}, \qquad \left\{ \sigma^2, \sigma^3 \right\} .$$

The class equation is

$$10 = 1 + 5 + 2 + 2 .$$

We see that in the case $n = 4$, the two different types of reflections each form a conjugacy class. When $n = 5$ there is only one type of reflection and only one conjugacy class. It is not hard to generalize this calculation to arbitrary n (see Exercise 9.2).   □

A First Application

Our first application is a result that is useful in classifying groups whose order is a prime power.

DEFINITION 9.1   *A p-group is a group of order p^s for some $s > 0$.*

For example, D_4 and Q are 2-groups, and $\mathbb{Z}/3\mathbb{Z} \times \mathbb{Z}/3\mathbb{Z}$ is a 3-group.

THEOREM 9.2
Suppose G is a p-group for some prime p. Then the centre of G is not trivial.

PROOF   According to the class equation

$$|G| = |Z(G)| + \sum_{j=1}^{r} |C_{\alpha_j}|$$

where $\alpha_1, \ldots, \alpha_s$ are representatives of the nontrivial conjugacy classes of G. Since the order of each nontrivial conjugacy class divides $|G|$, they must each be a power of p. Therefore their sum is a multiple of p. Hence p divides $|Z(G)|$ as well.   ∎

For example, as was shown in the previous section, and in Exercise 8.12, the centre of D_4 has order 2 and is thus not trivial. This theorem will allow us to classify groups of order p^2.

Burnside's Counting Lemma

Our second application is a formula for the number of orbits of a finite group acting on a finite set. It is useful in combinatorial problems with symmetry. First we need a definition. If H is a subgroup of a group G, the *conjugate* of H by $\alpha \in G$ is the subgroup

$$\alpha H \alpha^{-1} := \left\{ \alpha \beta \alpha^{-1} \mid \beta \in H \right\} .$$

Two subgroups H and K are *conjugate* to one another if there exists an $\alpha \in G$ such that

$$K = \alpha H \alpha^{-1} .$$

In Exercise 8.9 you saw that if G acts on X and two points $x, y \in X$ lie in the same orbit, then their stabilizers G_x and G_y are conjugate to one another.

THEOREM 9.3 (Burnside's Lemma)
Let G be a finite group acting on a finite set X. Denote by m_α, the number of fixed points of $\alpha \in G$ and by s, the number of orbits of G in X. Then

$$s = \frac{1}{|G|} \sum_{\alpha \in G} m_\alpha \,.$$

PROOF   First suppose that G acts transitively. So $s = 1$, and we want to show that

$$|G| = \sum_{\alpha \in G} m_\alpha \,.$$

Set
$$Y = \{(\alpha, x) \in G \times X \mid \alpha x = x\} \,.$$

Now we count $|Y|$ in two different ways. If we pick an $x \in X$, then $(\alpha, x) \in Y$ if and only if $\alpha \in G_x$. So the number of such pairs is $|G_x|$. For any $y \in X$, G_y is conjugate to G_x and therefore $|G_y| = |G_x|$. Hence, summing over y, we have

$$|Y| = \sum_{y \in X} |G_y| = \sum_{y \in X} |G_x| = |X||G_x| = |G| \,,$$

by Formula 9.1. On the other hand, if we choose an $\alpha \in G$, then the $x \in X$ such that $(\alpha, x) \in Y$ are just the fixed points of α. So summing over α, we get

$$|Y| = \sum_{\alpha \in G} m_\alpha \,.$$

We can now prove the general case. Since G acts transitively on each orbit, the formula we have just proved applies to each orbit. The total number of fixed points an element α has, is the sum of the number of fixed points in each orbit. Therefore

$$s|G| = \sum_{\alpha \in G} m_\alpha \,. \quad \blacksquare$$

In Example 8.2 the group V' acts on $\{1, 2, 3, 4\}$. There are 2 orbits. Let us count the fixed points. For $\alpha = (1)$, we have $m_\alpha = 4$. For α a transposition, $m_\alpha = 2$. For $\alpha = (1\,2)(3\,4)$, $m_\alpha = 0$. So Burnside's formula is

$$2 = \frac{1}{4}(4 + 2 + 2 + 0) \,.$$

Example 9.2
Suppose we want to count the number of ways of colouring the vertices of a regular pentagon black or white. Since there are five vertices, and two ways to colour each one, the simplest answer is:
$$2^5 = 32.$$

But suppose we are making a necklace with five beads, each coloured black or white. Then we do not want to distinguish between two patterns that can be transformed into one another by a symmetry of the pentagon, for example

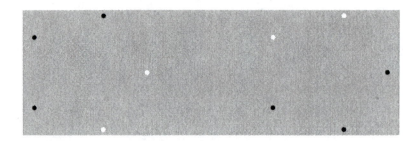

To count these, let X be the set of 32 patterns. The symmetry group of the pentagon, D_5, acts on X. We want to know how many orbits there are. The Burnside formula will tell us, once we have computed the number of fixed points of each symmetry. There are three different types of symmetries to consider. First, $\alpha = 1$. Then $m_\alpha = 32$. Secondly, α could be one of the four nontrivial rotations. The only patterns that a rotation leaves invariant are the two that are all black or all white. So in this case, $m_\alpha = 2$. Lastly, α could be one of the five reflections. Recall that these reflect the pentagon in a line passing through a vertex and the midpoint of the opposite side. So one vertex is fixed and the other four are interchanged in pairs. There are two ways of colouring each pair and of colouring the fixed vertex. So $m_\alpha = 2^3 = 8$. Substituting these numbers into the formula, we have

$$10s = 32 + 4 \cdot 2 + 5 \cdot 8 = 80 \,,$$

where s is the number of orbits. Therefore $s = 8$. Here are 8 patterns that represent the 8 orbits.

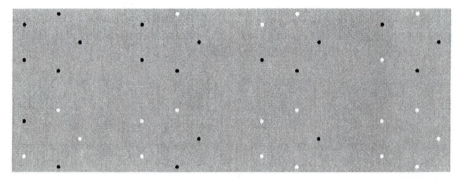

[]

Finite Subgroups of $SO(3)$

Our second application of formula (9.1) is to find the finite subgroups of $SO(3)$. We already know the subgroups $\mathbb{T}, \mathbb{O}, \mathbb{I}$. The dihedral groups also can be realized as groups of rotations of geometric objects. Take a regular n-gon in the plane. Construct a pyramid above it and one of the same height below it. A rotation of the n-gon in the plane can be extended to a rotation of the solid about the line joining the peaks of the two pyramids. A reflection can be extended to a rotation through an angle of π about the axis of the reflection. Thus D_n can be embedded in $SO(3)$ as a group of symmetries of this solid. Since D_n contains a cyclic subgroup of order n, it too is a subgroup of $SO(3)$. We shall see that these are essentially all the finite subgroups. The way we shall demonstrate this is to consider the fixed points of a group of rotations acting on the unit sphere S^2,

$$S^2 := \left\{ v \in \mathbb{R}^3 \mid \|v\| = 1 \right\}.$$

REMARK 9.2   If $\alpha \in O(3)$, then for any $v \in \mathbb{R}^3$, $\|\alpha v\| = \|v\|$, in particular if $\|v\| = 1$, then $\|\alpha v\| = 1$. So $O(3)$ acts on S^2. Any subgroup of $O(3)$, for example $\mathbb{T}$, also acts on S^2.   ∎

DEFINITION 9.2   *If a group G acts on a set X, the set of fixed points of G is*

$$\left\{ x \mid \alpha x = x \text{ for some } \alpha \neq 1 \right\} = \left\{ x \mid G_x \neq \{1\} \right\}.$$

For example, take $G = \mathbb{T}$, acting on S^2. Each nontrivial element in $\mathbb{T}$ is a rotation about an axis. The axis meets S^2 in a pair of antipodal points, which are fixed by the rotation. These two points belong to the set of fixed points. The rotations about a line through a vertex and the centre of the opposite face give four pairs of fixed points. The rotations about an axis joining the midpoints of a pair of opposite edges give another three pairs. In the following picture, the arcs on the sphere are the edges of an inscribed tetrahedron projected onto the sphere. The centre of one face is shown.

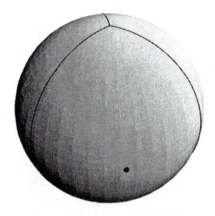

THEOREM 9.4

Let $G < SO(3)$ be a finite subgroup. Then G is conjugate to a cyclic group, to D_n, $n \geq 2$, to $\mathbb{T}$, to $\mathbb{O}$, or to $\mathbb{I}$.

PROOF   As a subgroup of $SO(3)$, G acts on S^2. Each nontrivial rotation in G fixes the two points where its axis meets the sphere. The set of all such pairs of antipodal points is the set of fixed points of G, which we shall denote by X. Now suppose $x \in X$ is fixed by $\alpha \in G$. Take any $\beta \in G$. Then βx is fixed by $\beta \alpha \beta^{-1}$. So $\beta x \in X$. Thus G acts on X.

Let $O_1, \ldots, O_s$ be the orbits of G in X. The stabilizer of a point in an orbit O_j has order

$$n_j := |G|/|O_j|, \tag{9.1}$$

for $1 \leq j \leq s$, by Formula 9.1. Since all the points in X have nontrivial stabilizers, $n_j \geq 2$.

We now count fixed points as in the proof of Burnside's Lemma. Let

$$Y = \{(\alpha, x) \mid \alpha x = x, \ \alpha \in G \setminus \{1\}, \ x \in X\} \, .$$

For fixed x, $(\alpha, x) \in Y$ if and only if $\alpha \in G_x \setminus \{1\}$. If $x \in O_j$, then the number of such elements α is $n_j - 1$. So the points $x \in O_j$ contribute $|O_j|(n_j - 1)$ elements

α. Summing over j, we then get

$$|Y| = \sum_{j=1}^{s} |O_j|(n_j - 1) .$$

On the other hand, if we fix $\alpha \in G \setminus \{1\}$, then $(\alpha, x) \in Y$ if and only if x is a fixed point of α. As we already noted, each rotation α has 2 fixed points. So summing over $\alpha \in G \setminus \{1\}$, we obtain

$$|Y| = \sum_{\alpha \in G \setminus \{1\}} 2 = 2(|G| - 1) .$$

Thus

$$2(|G| - 1) = \sum_{j=1}^{s} (n_j - 1)|O_j| .$$

Substitute the value of $|O_j|$ from Equation (9.1):

$$2(|G| - 1) = \sum_{j=1}^{s} \frac{n_j - 1}{n_j} |G| = |G|s - |G| \sum_{j=1}^{s} \frac{1}{n_j} .$$

Now divide through by $|G|$ and rearrange terms:

$$\sum_{j=1}^{s} \frac{1}{n_j} = s - 2 + \frac{2}{|G|} . \tag{9.2}$$

This is the equation we must analyze. First, notice that since all $n_j \geq 2$, each term on the left is at most $1/2$. So we have the inequality

$$\frac{s}{2} \geq s - 2 + \frac{2}{|G|} > s - 2 ,$$

which implies that

$$2 > \frac{s}{2} .$$

Thus $s \leq 3$. This leaves us with three cases to discuss.

(i) s = 1.

Equation (9.2) becomes:

$$\frac{1}{n_1} = \frac{2}{|G|} - 1 \leq 0 ,$$

since $|G| \geq 2$. Then $n_1 \leq 0$, which is impossible.

(ii) s = 2.

Equation (9.2) becomes:

$$\frac{1}{n_1} + \frac{1}{n_2} = \frac{2}{|G|} .$$

Multiplying by $|G|$, and inserting (9.1), we have

$$|O_1| + |O_2| = 2 .$$

Therefore $|O_1| = |O_2| = 1$ and $n_1 = n_2 = |G|$. Now if G has only two fixed points, they must be antipodal. And the line passing through them must be the axis of rotation of the elements of G. These are then rotations in the plane perpendicular to this axis. So G can be regarded as a subgroup of $SO(2)$. We saw earlier that finite subgroups of $SO(2)$ are cyclic. Therefore G is cyclic.

(iii) s = 3.

Equation (9.2) becomes:

$$\frac{1}{n_1} + \frac{1}{n_2} + \frac{1}{n_2} = 1 + \frac{2}{|G|} . \tag{9.3}$$

In particular

$$\frac{1}{n_1} + \frac{1}{n_2} + \frac{1}{n_3} > 1 .$$

LEMMA 9.1

The solutions of this inequality, with the constraints $n_1, n_2, n_3 \geq 2$ are

n_1	n_2	n_3	$|G|$
2	2	n	2n
2	3	3	12
2	3	4	24
2	3	5	60

PROOF   If $n_1, n_2, n_3 \geq 3$, then

$$\frac{1}{n_1} + \frac{1}{n_2} + \frac{1}{n_3} \leq 1 .$$

So at least one is 2, say $n_1 = 2$. If $n_2, n_3 \geq 4$, then

$$\frac{1}{2} + \frac{1}{n_2} + \frac{1}{n_3} \leq 1 .$$

Therefore, we can assume that $n_2 < 4$. The first possible solution is $n_1 = 2, n_2 = 2, n_3 = n$, where $n \geq 2$ is arbitrary. Now suppose $n_2 = 3$. Then the inequality becomes

$$\frac{1}{2} + \frac{1}{3} + \frac{1}{n_3} > 1 .$$

Thus we must have that $n_3 < 6$. This gives the other three solutions in the table. To compute $|G|$ substitute the values of n_1, n_2, and n_3 in Equation (9.3). ∎

We return to the proof of the theorem. The entries in the table correspond to D_n, $\mathbb{T}$, $\mathbb{O}$, and $\mathbb{I}$, respectively. The four cases are similar. We will do the second one.

Suppose that $n_1 = 2, n_2 = 3, n_3 = 3$, and $|G| = 12$. Then $|O_1| = 6, |O_2| = 4$, and $|O_3| = 4$. We want to prove that these orbits are the set of midpoints of the edges of a regular tetrahedron, the set of its vertices, and the set of centres of its faces. Begin with a point in $P_1 \in O_2$. Let l_1 be the line through P_1 and the origin. It is the axis of the rotations in G_{P_1}, which have angles of rotation $2\pi/3$ and $4\pi/3$. Pick a point $P_2 \in O_2$ different from P_1. Its orbit under G_{P_1} is $\{P_2, P_3, P_4\}$. They all lie in a plane perpendicular to l_1 and are the vertices of an equilateral triangle in this plane.

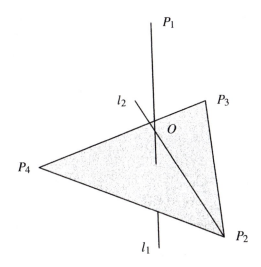

Take the line joining any of these to the origin, say the line l_2 joining P_2 to the origin. It is the axis for the rotations of G_{P_2}, which permute $\{P_1, P_3, P_4\}$. So these all lie in a plane perpendicular to l_2 and form an equilateral triangle. Thus the 4 points in O_2 are equidistant from one another and are the vertices of a regular tetrahedron. The group G is the group of proper symmetries of this tetrahedron. The orbits O_3 and O_1 are the centres of its faces and the midpoints of its edges, respectively. The symmetry groups of any two regular tetrahedra are conjugate in $SO(3)$. This proves the second case. ∎

This result can also be proved using only spherical geometry: see [8, § 3.8].

Exercises

9.1 Verify the class formula of S_5.

9.2 Determine the conjugacy classes of D_n, for $n \geq 4$.

9.3 Calculate the terms in the class equation for $SL(2, \mathbb{F}_5)$.

9.4 Calculate the terms in the class equation of A_6.

9.5 Use the *Mathematica* function `ConjugacyClass` to compute the conjugacy classes of A_6.

9.6 What is the centre of the permutation group of degree 8 generated by $\{(1\ 2\ 3\ 4\ 5\ 6\ 7\ 8), (1\ 3\ 5\ 7)\}$?

9.7 How many different necklaces with six beads can be made from beads of three colours?

9.8 How many ways can the faces of a cube be coloured black and white?

9.9 Complete the proof of Theorem 9.4 in the case of the cube.

9.10 Show that $D_6 \cong S_3 \times \mathbb{Z}/2\mathbb{Z}$.

9.11 Let G be a group of order 10. Let X be the set of pairs of elements of G:

$$X = \left\{ \{\xi, \nu\} \mid \xi, \nu \in G, \xi \neq \nu \right\}.$$

Then G acts on X by left multiplication:

$$\alpha \cdot \{\xi, \nu\} = \{\alpha\xi, \alpha\nu\}.$$

   (i) What is $|X|$? Show that there is an orbit of length 5.
   (ii) Conclude that G has a subgroup of order 2.
   (iii) Give an analogous argument to show that G has a subgroup of order 5.
   (iv) Suppose that H is a group of order 20. Prove that H has a subgroup of order 4.

Chapter 10

Cosets

Lagrange's Theorem

At the beginning of the previous chapter when we looked at the evaluation map $e : G \to O_x$, we came upon subsets of G of the form αG_x, where $x \in X$, $\alpha \in G$. Such subsets are called cosets of the stabilizer G_x in G. In general, given a subgroup $K < G$, we call a subset

$$\alpha K := \{\alpha\kappa \mid \kappa \in K\}$$

a *left coset* of K in G. For example, a left coset of $n\mathbb{Z}$ in $\mathbb{Z}$ is a set of the form $m + n\mathbb{Z}, m \in \mathbb{Z}$. This is just the congruence class of m modulo n. A left coset is not a subgroup of G except for the one coset $1 \cdot K = K$, because this is the only coset containing 1.

Examples 10.1

(i) What are the left cosets of A_n in S_n? Well, if $\alpha \in S_n$ is even, then $\alpha A_n = A_n$. If α is odd, then αA_n is the set of odd permutations. So there are two cosets: the set of even permutations and the set of odd permutations.

(ii) Suppose we take $G = S_3$ and $K = \langle (1\,2) \rangle$. Then

$$(1)K = K$$
$$(1\,2)K = K$$
$$(1\,3)K = \{(1\,3), (1\,3\,2)\}$$
$$(2\,3)K = \{(2\,3), (1\,2\,3)\}$$
$$(1\,2\,3)K = \{(2\,3), (1\,2\,3)\}$$
$$(1\,3\,2)K = \{(2\,3), (1\,2\,3)\}.$$

Thus there are three left cosets: $K = \{(1), (1\,2)\}$, $\{(1\,3), (1\,3\,2)\}$, and $\{(2\,3), (1\,2\,3)\}$.

(iii) Here are the left cosets of V in A_4:

$$V = \{(1), \ (1\,2)(3\,4), \ (1\,3)(2\,4), \ (1\,4)(2\,3)\}$$
$$(1\,2\,3)V = \{(1\,2\,3), \ (2\,4\,3), \ (1\,4\,2), \ (1\,3\,4)\}$$
$$(1\,3\,2)V = \{(1\,3\,2), \ (1\,4\,3), \ (2\,3\,4), \ (1\,2\,4)\} \ .$$

(Check this calculation yourself!)

(iv) Let $G = \mathbb{R}^2$ with vector addition, and let K be a line through $(0, 0)$. The left cosets of K in G are the translates $v + K$, $v \in \mathbb{R}^2$, of K.

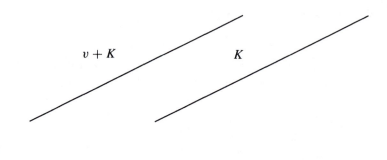

□

The last section of this chapter explains *Mathematica* functions that compute cosets. One can also look at the *right cosets* of K in G: they are subsets of the form $K\alpha$, $\alpha \in G$. We denote the set of left cosets of K in G by G/K. We already used the notation $\mathbb{Z}/n\mathbb{Z}$ for the set of congruence classes mod n, otherwise known as the integers mod n. If G/K is finite, then the number of elements in it is called the *index* of K in G, written $[G : K]$.

Notice that in the examples we have just looked at, every left coset has the same number of elements, namely $|K|$, and distinct cosets are disjoint. As a result of this, $[G : K]|K| = |G|$. For example, there are three left cosets of V in A_4. Each has four elements, and $|A_4| = 12$.

THEOREM 10.1 (Lagrange's Theorem)
If G is a finite group and K a subgroup of G, then

$$|G| = [G : K]|K| \ .$$

PROOF   The theorem can be proved in the way suggested above. This is done in Exercise 10.1. It also follows from Formula 9.1, as we are going to see now. The group G acts on the set G/K by left multiplication: define

$$\alpha \cdot (\beta K) := (\alpha\beta)K \ ,$$

for $\alpha, \beta \in G$. This is a variant of the action of G on itself by multiplication on the left and you show that it is an action in the same way. It too is transitive: given two cosets, $\beta K, \gamma K \in G/K$, the group element $\alpha = \gamma \beta^{-1}$ satisfies

$$\alpha \cdot \beta K = \gamma K .$$

So there is only the one orbit, with $[G : K]$ points in it. The stabilizer of the coset K is the subgroup K. We can now apply our formula relating the number of points in an orbit to the order of the stabilizer:

$$|G| = [G : K]|K| . \quad \blacksquare$$

COROLLARY 10.1
The order of a subgroup divides the order of the group.

For example, S_3 can have subgroups of order 1, 2, and 3, but not of order 4 or 5. It is not hard to write down these subgroups. First, there is the trivial subgroup of order 1. Subgroups of order 2 are those generated by a transposition. There are three of these. There is exactly one subgroup of order 3, namely $A_3 = \langle(123)\rangle$. The graph below shows how these subgroups fit together. It is called the lattice of subgroups of S_3.

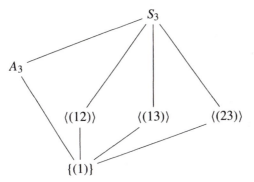

COROLLARY 10.2
The order of an element divides the order of the group.

A consequence of this is that if $|G| = n$, then for any $\alpha \in G$

$$\alpha^n = 1 .$$

COROLLARY 10.3
A group of prime order is cyclic.

PROOF   Let G be a group of order p, where p is prime. The order of any element in G must divide p. Therefore it must be either 1 or p. So if $\alpha \in G, \alpha \neq 1$, then $\langle \alpha \rangle = G$.   ■

Thus groups of order 2, 3, 5, and 7 are all cyclic. On the other hand, we know of a group of order 4 that is not cyclic, namely V, and one of order 6 that is not cyclic, S_3.

Let us apply Corollary 10.2 to the group $\mathbb{F}_p^\times$. Since it has order $p - 1$, the corollary tells us that $a^{p-1} = 1$ for any $a \in \mathbb{F}_p^\times$, or equivalently

$$a^{p-1} \equiv 1 \pmod{p}$$

for any $a \in \mathbb{Z}, (a, p) = 1$. But we can extend this to all integers if we multiply the congruence by a:

THEOREM 10.2 (Fermat's Little Theorem)

$$a^p \equiv a \pmod{p}$$

for all $a \in \mathbb{Z}$.

The converse to Lagrange's theorem is not true. If G is a finite group and d divides $|G|$, there need not be a subgroup of order d. Here is an example.

Example 10.2
Look at A_4. It consists of eight 3-cycles, three products of disjoint transpositions, and the identity. Each 3-cycle generates a cyclic subgroup of order 3. Any two of them generate the whole group. Each element of type $\{2, 2\}$ generates a cyclic subgroup of order 2. Two of them generate the subgroup V of order 4. A 3-cycle and a product of two transpositions generate the whole group. So there is no subgroup of order 6. Here is the lattice of subgroups of A_4 .

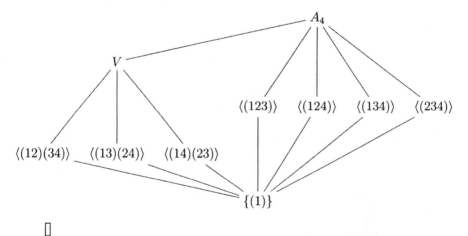

▯

Normal Subgroups

Another situation where cosets naturally arise is the following. Suppose $f : G \to H$ is a group homomorphism. When do two elements in G have the same image in H? Well, let $\alpha, \alpha' \in G$, $\bar{\alpha} \in H$ with

$$f(\alpha) = f(\alpha') = \bar{\alpha} .$$

Then

$$f(\alpha^{-1}\alpha') = 1_H$$

so that

$$\alpha^{-1}\alpha' \in \ker(f)$$

or

$$\alpha' \in \alpha \ker(f) .$$

Conversely, if $f(\alpha) = \bar{\alpha}$ and $\alpha' \in \alpha \ker(f)$, then $f(\alpha') = \bar{\alpha}$. So

$$f^{-1}(\bar{\alpha}) = \alpha \ker(f) .$$

The kernel and its cosets have a special property. The most convenient way to express it is this: for any $\alpha \in G$,

$$\alpha \ker(f)\alpha^{-1} = \ker(f) .$$

(For $\beta \in \ker f$, $f(\alpha\beta\alpha^{-1}) = f(\alpha)f(\alpha^{-1}) = 1_H$). We give a name to subgroups with this property.

DEFINITION 10.1   *A subgroup K of G is called a normal subgroup, written $K \lhd G$, if*

$$\alpha K \alpha^{-1} = K ,$$

for all $\alpha \in G$.

This property can be expressed in terms of the cosets of K.

THEOREM 10.3
A subgroup K of a group G is normal if and only if

(i) for any $\alpha \in G$, $\alpha K = K\alpha$, or equivalently,

(ii) for any $\alpha, \beta \in G$, $(\alpha K)(\beta K) = (\alpha\beta)K$, or equivalently,

(iii) for any $\alpha \in G$, $\alpha K \alpha^{-1} \subset K$.

PROOF   Suppose that K is a normal subgroup. Take an $\alpha \in G$. We have that

$$\alpha K \alpha^{-1} = K .$$ (10.1)

Multiplying on the right by α we get

$$\alpha K = K \alpha .$$

Thus every left coset coincides with the corresponding right coset. In terms of elements of K, this means that for $\kappa \in K$, there exists a $\lambda \in K$, such that

$$\alpha \kappa = \lambda \alpha ,$$

and vice versa. Now assume that Equation (i) holds for all α. Then given $\alpha, \beta \in G$,

$$(\alpha K)(\beta K) = \alpha(K\beta)K = \alpha(\beta K)K = (\alpha\beta)K$$

So the product of the cosets of α and of β is the coset of $\alpha\beta$. If Equation (ii) holds, then taking $\beta = \alpha^{-1}$, we have

$$(\alpha K)(\alpha^{-1} K) = (\alpha\alpha^{-1})K = 1 \cdot K = K ,$$ (10.2)

which implies that

$$\alpha K \alpha^{-1} \subset K .$$ (10.3)

(Why? Well, lying in the set on the left-hand side of (10.2) are all elements of the form $(\alpha\kappa)(\alpha^{-1}1)$, with $\kappa \in K$). Finally, since Equation (10.3) holds for all elements of G, it holds for α^{-1}:

$$\alpha^{-1} K \alpha \subset K .$$

Conjugating both sides by α gives

$$K \subset \alpha K \alpha^{-1} ,$$

and combining this with (10.3) gives Equation (10.1).   ∎

   In practice, to check whether $K \lhd G$, you need only verify whether $\alpha K \alpha^{-1} \subset K$, when α runs through a set of generators of G. In fact it is sufficient to check if $\alpha\kappa\alpha^{-1} \in K$, where α runs through a set of generators of G and κ runs through a set of generators of K. In Example 10.1(iii) we know that $\{(1\ 2\ 3),\ (1\ 2)(3\ 4)\}$ generates A_4. Since $(1\ 2)(3\ 4) \in V$, it is enough to check that $(1\ 2\ 3)V(1\ 3\ 2) = V$. We see then that V is a normal subgroup of A_4.

REMARK 10.1   Any subgroup K of index 2 is normal. There are two left cosets: K and αK, where $\alpha \notin K$. And there are two right cosets: K and $K\alpha$. The left cosets are disjoint from one another and so are the right cosets. Therefore $\alpha K = K\alpha$, and K is normal. For example, A_n is a normal subgroup of S_n, for all n. ∎

If G is abelian, then every subgroup is normal since $\alpha\beta\alpha^{-1} = \beta$ for any α and β. The centre of any group is a normal subgroup because the same equation holds for any α, with β in the centre.

We should also see an example of a subgroup that is not normal. Let $G = S_3$ and $K = \langle (1\ 2) \rangle$. Then $(1\ 2\ 3)(1\ 2)(1\ 3\ 2) = (2\ 3) \notin K$. So K is not normal.

Quotient Groups

Equation (ii) in Theorem 10.3 tells us that we can make G/K into a group by multiplying cosets when K is a normal subgroup (and only then). Let us do this carefully. First, we have a binary operation on G/K:

$$G/K \times G/K \to G/K \ ,$$

given by

$$(\alpha K, \beta K) \mapsto (\alpha K)(\beta K) = \alpha\beta K \ ,$$

for $\alpha, \beta \in G$. This operation is associative:

$$(\alpha K \beta K)\gamma K = (\alpha\beta)K\gamma K = (\alpha\beta)\gamma K = \alpha(\beta\gamma)K = \alpha K(\beta\gamma)K = \alpha K(\beta K \gamma K) \ .$$

Secondly, there is an identity element, namely $1 \cdot K = K$:

$$(\alpha K)K = \alpha K = K(\alpha K) \ .$$

Thirdly, the inverse of a coset αK is $\alpha^{-1}K$:

$$\alpha K \alpha^{-1} K = (\alpha\alpha^{-1})K = K = \alpha^{-1}K\alpha K \ .$$

G/K with this operation is called the *quotient group* of G mod K.

You have already seen a quotient group: $\mathbb{Z}/n\mathbb{Z}$, the integers mod n, which is the quotient group of the subgroup $n\mathbb{Z}$. In fact, historically this is the example that lead to the general construction. If we look again at Example 10.1(iii), we see that

$$\big((1\ 2\ 3)V\big)^2 = (1\ 2\ 3)V(1\ 2\ 3)V = (1\ 2\ 3)^2V = (1\ 3\ 2)V$$

$$\big((1\ 2\ 3)V\big)^3 = (1\ 2\ 3)V(1\ 3\ 2)V = V \ .$$

This shows that A_4/V is a cyclic group of order 3.

Using *Mathematica* to generate left cosets and then multiply them together makes it easy to see the multiplication in G/K in examples where $[G : K]$ is larger.

The Canonical Isomorphism

We noted at the beginning of our discussion of normal subgroups that the kernel of a homomorphism is normal. In fact every normal subgroup is the kernel of a homomorphism. For example, let K be a normal subgroup of a group G. We have a canonical map

$$p : G \twoheadrightarrow G/K$$

given by

$$p : \alpha \mapsto \alpha K \ ,$$

where $\alpha \in G$. By the definition of the group operation in G/K, this map is a homomorphism:

$$p(\alpha\beta) = \alpha\beta K = \alpha K \beta K = p(\alpha)p(\beta)$$

for $\alpha, \beta \in G$. It is surjective, and

$$p(\alpha) = 1_{G/K} = K \qquad \text{if and only if} \qquad \alpha \in K$$

Thus the kernel of p is K.

Example 10.3

Take $G = S_4$ and $K = V$. It is easy to check that V is a normal subgroup of S_4. The quotient group S_4/V has order $24/4 = 6$. We think we know all groups of order 6. Which one is it? Three of the cosets are written out in Example 10.1(iii). The other three are

$$(1\,2)V = \{(1\,2), \ (3\,4), \ (1\,3\,2\,4), \ (1\,4\,2\,3)\}$$
$$(1\,3)V = \{(1\,3), \ (2\,4), \ (1\,2\,3\,4), \ (1\,4\,3\,2)\}$$
$$(2\,3)V = \{(2\,3), \ (1\,4), \ (1\,2\,4\,3), \ (1\,3\,4\,2)\}$$

Thus the six cosets can be written as

$$V, \ (1\,2)V, \ (1\,3)V, \ (2\,3)V, \ (1\,2\,3)V, \ (1\,3\,2)V \ .$$

We can define a map

$$S_3 \twoheadrightarrow S_4/V$$

by

$$\alpha \mapsto \alpha V \ ,$$

for $\alpha \in S_3$. Because of the definition of the group operation in S_4/V, this map is a homomorphism. Since it is bijective, we have

$$S_4/V \cong S_3 \ .$$

Here is a more sophisticated way of presenting the same argument. Consider the subgroup H of S_4,

$$H := \{(1),\ (1\,2),\ (1\,3),\ (2\,3),\ (1\,2\,3),\ (1\,3\,2)\} \cong S_3 .$$

Then $H \cap V = \{(1)\}$. So the canonical homomorphism $p : S_4 \to S_4/V$ is one-to-one on H. Because both H and S_4/V have order 6, they are in fact isomorphic and

$$S_4/V \cong S_3 . \quad \square$$

In Example 10.5 below we will look at a more complicated example.

In a sense, every homomorphism looks like the canonical homomorphism. This is what the following theorem says.

THEOREM 10.4 (First Isomorphism Theorem)
Let $f : G \to H$ be a homomorphism of groups. Then $f = \bar{f} p$ where

$$p : G \to G/\ker(f)$$

is the canonical homomorphism and

$$\bar{f} : G/\ker(f) \cong \operatorname{im}(f) .$$

PROOF As we saw earlier, for any $\alpha' \in \alpha \ker(f) \subset G$

$$f(\alpha') = f(\alpha) .$$

Therefore we get a well-defined mapping

$$\bar{f} : G/\ker(f) \to H$$

if we set

$$\bar{f}\big(\alpha \ker(f)\big) := f(\alpha) .$$

This mapping is a homomorphism because f is one:

$$\bar{f}\big(\alpha \ker(f)\beta \ker(f)\big) = \bar{f}\big(\alpha\beta \ker(f)\big) = f(\alpha\beta) = f(\alpha)f(\beta)$$
$$= \bar{f}\big(\alpha \ker(f)\big)\bar{f}\big(\beta \ker(f)\big) .$$

As noted above as well, $\bar{f}$ is injective: for $\bar{\alpha} = f(\alpha) \in \operatorname{im}(f)$,

$$f^{-1}(\bar{\alpha}) = \alpha \ker(f) .$$

And the image of $\bar{f}$ is just the image of f. So

$$\bar{f} : G/\ker(f) \xrightarrow{\cong} \operatorname{im}(f) .$$

and $f = \bar{f}p$. Here is a diagram.

$$
\begin{array}{ccc}
G & \xrightarrow{\ \ f\ \ } & H \\
\ \downarrow{\scriptstyle p} & & \uparrow \\
G/(\ker f) & \xrightarrow[\cong]{\ \ \bar{f}\ \ } & \text{im } f
\end{array}
$$

∎

Examples 10.4

(i) In Exercise 5.11 you looked at the exponential homomorphism $\exp : \mathbb{R} \to S$ given by
$$
\exp(x) = e^{2\pi i x} .
$$
Its kernel is $\mathbb{Z}$ and it is surjective. Therefore it induces an isomorphism
$$
\overline{\exp} : \mathbb{R}/\mathbb{Z} \xrightarrow{\cong} S .
$$

(ii) For any field F we have the homomorphism
$$
\det : GL(2, F) \to F^{\times} .
$$
By definition, its kernel is $SL(2, F)$ and it is surjective. Thus we have an isomorphism
$$
GL(2, F)/SL(2, F) \cong F^{\times} .
$$
This holds in particular for $F = \mathbb{F}_p$. We know that the order of $GL(2, \mathbb{F}_p)$ is $(p-1)^2 p(p+1)$. So we can compute $|SL(2, \mathbb{F}_p)|$:
$$
p - 1 = |\mathbb{F}_p^{\times}| = |GL(2, \mathbb{F}_p)|/|SL(2, \mathbb{F}_p)| = (p-1)^2 p(p+1)/|SL(2, \mathbb{F}_p)|
$$
and therefore
$$
|SL(2, \mathbb{F}_p)| = (p-1)p(p+1) .
$$

(iii) The action of $GL(2, \mathbb{F}_p)$ on $P(\mathbb{F}_p)$ gave us a homomorphism
$$
f_p : GL(2, \mathbb{F}_p) \to S_{p+1} .
$$
The kernel is $Z\big(GL(2, \mathbb{F}_p)\big) \cong \mathbb{F}_p^{\times}$. The map f_p then induces an injective homomorphism
$$
\bar{f}_p : PGL(2, \mathbb{F}_p) := GL(2, \mathbb{F}_p)/Z\big(GL(2, \mathbb{F}_p)\big) \to S_{p+1} .
$$
$PGL(2, \mathbb{F}_p)$ is called the projective linear group. Similarly one defines $PSL(2, \mathbb{F}_p)$, which we shall discuss further in Chapter 12.

⬚

Software and Calculations

For computing cosets there are functions `LeftCosets` and `RightCosets`. They take as arguments a group G and a subgroup K and produce a list of the elements of G partitioned into cosets with K itself as the first coset. To illustrate, let us repeat Example 10.1(iii). We have:

In[1]:= A4 = Group[P[{1,2,3}], P[{2,3,4}]]

Out[1]= ⟨(1 2 3), (2 3 4)⟩

and

In[2]:= V = Group[P[{1,2},{3,4}], P[{1,3},{2,4}]]

Out[2]= ⟨(1 2)(3 4), (1 3)(2 4)⟩

Then

In[3]:= LeftCosets[A4,V]

Out[3]= {{(1), (1 2)(3 4), (1 3)(4 2), (1 4)(2 3)},
{(1 2 3), (1 3 4), (1 4 2), (2 4 3)},
{(1 2 4), (1 3 2), (1 4 3), (2 3 4)}}

If you only want a representative from each coset you can use `LeftCosetReps` or `RightCosetReps`:

In[4]:= LeftCosetReps[A4,V]

Out[4]= {(1), (1 2 3), (1 2 4)}

These are just the first elements from each coset. We can check that V satisfies 10.3(i) by computing its right cosets and comparing them with the left cosets:

In[5]:= RightCosets[A4,V]

Out[5]= {{(1), (1 2)(3 4), (1 3)(2 4), (1 4)(2 3)},
{(1 3 2), (1 4 3), (1 2 4), (2 3 4)},
{(1 4 2), (1 2 3), (1 3 4), (2 4 3)}}

So you can multiply two cosets together to get a third:

```
In[6]:= {P[{1, 2, 3}], P[{1, 3, 4}], P[{1, 4, 2}],
         P[{2, 4, 3}]}.
        {P[{1, 2, 4}], P[{1, 3, 2}], P[{1, 4, 3}],
         P[{2, 3, 4}]}

Out[6]= {(1), (1 2)(3 4), (1 3)(2 4), (1 4)(2 3)}
```

We can also verify that V is normal in A_4 by calculating its conjugates. First we have the function Conjugate[a,b] which conjugates b by a. For example, if

```
In[7]:= a = P[{1,2,3}]

Out[7]= (1 2 3)
```

```
In[8]:= b = P[{1,4}]

Out[8]= (1 4)
```

then

```
In[9]:= Conjugate[a,b]

Out[9]= (2 4)
```

You can conjugate all the elements of a set by a at once:

```
In[10]:= Conjugate[ a, Elements[V] ]

Out[10]= {(1), (1 4)(2 3), (1 2)(3 4), (1 3)(2 4)}
```

that confirms V is normal in A_4.

Example 10.5

In S_5, we have the subgroup F_{20} (see Exercise 8.19):

```
In[11]:= F20 = Group[ P[{1,2,3,4,5}] , P[{1,2,4,3}] ]

Out[11]= ⟨(1 2 3 4 5), (1 2 4 3)⟩
```

A pair of generators of S_5 is

```
In[12]:= a = P[{1,2,3,4,5}]
         b = P[{1,2}]
```

Out [12] = (1 2 3 4 5)

Out [13] = (1 2)

To check whether F_{20} is normal in S_5, you conjugate the generators of F_{20} first by a and then by b, and look whether the resulting sets lie in F_{20}. In fact since a belongs to F_{20} you need only check

In [14] := Conjugate [b, Generators [F20]]

Out [14] = { (1 3 4 5 2), (1 4 3 2) }

which does not lie in F_{20}. So F_{20} is not normal. ▯

Example 10.6
Let G_{72} (see Exercise 8.16) be the permutation group

In [15] := G72 = Group [P[{1,2,3}], P[{1,4},{2,5},{3,6}],
                    P[{1,5,2,4},{3,6}]]

Out [15] = ⟨(1 2 3), (1 4)(2 5)(3 6), (1 5 2 4)(3 6)⟩

G_{72} has order 72:

In [16] := Order [G72]

Out [16] = 72

Let K be the subgroup

In [17] := K = Group [P[{1,2,3}], P[{4,5,6}]]

Out [17] = ⟨(1 2 3), (4 5 6)⟩

Since these two 3-cycles commute with one another, $K \cong \mathbb{Z}/3\mathbb{Z} \times \mathbb{Z}/3\mathbb{Z}$. The first element of the list of generators of G_{72} is also a generator of K. So to check that $K \triangleleft G_{72}$ we need only conjugate the generators of K by the remaining two generators of G_{72} :

In [18] := Conjugate [P[{1,4},{2,5},{3,6}],
            Generators [K]]
            Conjugate [P[{1,5,2,4},{3,6}],
            Generators [K]]

Out [18] = $\{(4\ 5\ 6),\ (1\ 2\ 3)\}$

Out [19] = $\{(4\ 5\ 6),\ (1\ 2\ 3)\}$

Therefore K is a normal subgroup of G_{72}. The quotient group $L := G_{72}/K$ has order $72/9 = 8$. We want to determine which group of order 8 it is. Since each coset has 9 elements, we will tell *Mathematica* not to print out the entire list of cosets when it computes L:

In [20] := L = LeftCosets [G72, K];

First we look at representatives from each of the cosets.

In [21] := LeftCosetReps [G72, K]

Out [21] = $\{(1),\ (1\ 2),\ (4\ 5),\ (1\ 4\ 2\ 5\ 3\ 6),$
$(1\ 4\ 2\ 6\ 3\ 5),\ (1\ 2)(4\ 5),$
$(1\ 4)(2\ 5\ 3\ 6),\ (1\ 4)(2\ 6\ 3\ 5)\}$

The first coset, $L[[1]]$, is $K = 1_L$. The second, $L[[2]] = (1\ 2)K$. Now

$$\left((1\ 2)K\right)^2 = (1\ 2)^2 K = K\ .$$

In other words, it has order 2. Since the representatives of the cosets $L[[3]]$ and $L[[6]]$ have order 2 as well, so do the cosets themselves. Now

In [22] := L[[4]].L[[4]]

Out [22] = $\{(1),\ (1\ 2\ 3),\ (1\ 3\ 2),\ (4\ 5\ 6),\ (4\ 6\ 5),$
$(1\ 2\ 3)(4\ 5\ 6),\ (1\ 2\ 3)(4\ 6\ 5),$
$(1\ 3\ 2)(4\ 5\ 6),\ (1\ 3\ 2)(4\ 6\ 5)\}$

which is just $L[[1]]$, and

In [23] := L[[5]].L[[5]]

Out [23] = $\{(1),\ (1\ 2\ 3),\ (1\ 3\ 2),\ (4\ 5\ 6),\ (4\ 6\ 5),$
$(1\ 2\ 3)(4\ 5\ 6),\ (1\ 2\ 3)(4\ 6\ 5),$
$(1\ 3\ 2)(4\ 5\ 6),\ (1\ 3\ 2)(4\ 6\ 5)\}$

as well. So $L[[4]]$ and $L[[5]]$ also have order 2. This leaves $L[[7]]$ and $L[[8]]$. We can see that

$$\left((1\ 4)(2\ 5\ 3\ 6)\right)^{-1} = (1\ 4)(2\ 6\ 3\ 5)\ .$$

So these two are inverse to each other. Do they in fact have order 4?

```
In[24]:= L[[7]].L[[7]]
```

Out[24]= { (1 2) (4 5), (1 2) (4 6), (1 2) (5 6),
           (1 3) (4 5), (1 3) (4 6), (1 3) (5 6),
           (2 3) (4 5), (2 3) (4 6), (2 3) (5 6 }

Comparing this with our list of coset representatives, we see that this must be $L[[6]]$, which has order 2. So $L[[6]] = L[[7]]^2$ and $L[[8]] = L[[7]]^3$. It begins to look as if L might be isomorphic to D_4 (see Equation (7.1)). To check this, we need generators σ and τ satisfying

$$\sigma^4 = 1, \quad \tau^2 = 1, \quad \sigma\tau = \tau\sigma^{-1}.$$

Let us try $\sigma = L[[7]]$ and $\tau = L[[2]]$. To see that they generate L we compute

```
In[25]:= L[[2]].L[[7]]
         L[[2]].L[[6]]
         L[[2]].L[[8]]
```

Out[25]= { (1 4 2 5 3 6), (1 4 3 6 2 5), (1 5 2 6 3 4),
           (1 5 3 4 2 6), (1 6 2 4 3 5), (1 6 3 5 2 4),
           (1 4) (2 5) (3 6), (1 5) (2 6) (3 4),
           (1 6) (2 4) (3 5) }

Out[26]= { (4 5), (4 6), (5 6), (1 2 3) (4 5),
           (1 2 3) (4 6), (1 2 3) (5 6), (1 3 2) (4 5),
           (1 3 2) (4 6), (1 3 2) (5 6) }

Out[27]= { (1 4 2 6 3 5), (1 4 3 5 2 6), (1 5 2 4 3 6),
           (1 5 3 6 2 4), (1 6 2 5 3 4), (1 6 3 4 2 5),
           (1 4) (2 6) (3 5), (1 5) (2 4) (3 6),
           (1 6) (2 5) (3 4) }
```

Thus

$$L[[2]].L[[7]] = L[[4]]$$
$$L[[2]].L[[7]]^2 = L[[2]].L[[6]] = L[[3]]$$
$$L[[2]].L[[7]]^3 = L[[2]].L[[8]] = L[[5]].$$

So $L[[7]]$ and $L[[2]]$ do generate $L$. Now let us see whether they satisfy the right relation. We must check whether

$$L[[7]]^{-1}.L[[2]] = L[[8]].L[[2]] = L[[4]] :$$
```

```
In [28] := L[[8]].L[[2]
```

```
Out [28] = {(1 4 2 5 3 6), (1 4 3 6 2 5), (1 5 2 6 3 4),
           (1 5 3 4 2 6), (1 6 2 4 3 5), (1 6 3 5 2 4),
           (1 4)(2 5)(3 6), (1 5)(2 6)(3 4),
           (1 6)(2 4)(3 5)},
```

which is $L[[4]]$. So indeed

$$L \cong D_4 .$$

◻

Exercises

10.1 Suppose that G is a finite group and $K < G$.

(a) Define a relation in G by

$$\alpha \sim \beta \quad \text{if} \quad \alpha K = \beta K .$$

Verify that this is an equivalence relation. Conclude that two cosets are either equal or disjoint.

(b) Show that $|\alpha K| = |K|$ for any $\alpha \in G$.

(c) From (a) and (b), deduce Lagrange's Theorem.

10.2 Prove that for natural numbers a and n, which are relatively prime,

$$a^{\varphi(n)} \equiv 1 \quad (\text{mod } n) .$$

10.3 • Draw the lattice of subgroups of D_4. Do the same for the quaternion group Q (see Exercise 4.5).

10.4 Which subgroups of D_4 are normal? Identify the corresponding quotient groups. Do the same for Q.

10.5 Verify that V is a normal subgroup of S_4. Find all normal subgroups of S_4.

10.6 Is the permutation group of order 72 in Example 10.6 a normal subgroup of S_6? Is $N(p)$ a normal subgroup of $GL(2, \mathbb{F}_p)$ (see Example 4.1(ii))?

10.7 Let G be a group and K a subgroup. Suppose that g is a set of generators of G and k of K. Show that if

$$\alpha \kappa \alpha^{-1} \in K \ ,$$

for all $\alpha \in g$ and $\kappa \in k$, then K is a normal subgroup.

10.8 • Let H and K be normal subgroups of a group G with $H \cap K = \{1\}$. Prove that $\alpha\beta = \beta\alpha$ for any $\alpha \in H$ and $\beta \in K$. Suggestion: show that $\alpha\beta\alpha^{-1}\beta^{-1} \in H \cap K$.

10.9 Check that in Example 10.1(ii) the product of two left cosets may not be a left coset.

10.10 Prove that a quotient group of a cyclic group is cyclic.

10.11 Show that $S_n/A_n \cong \mathbb{Z}/2\mathbb{Z}$.

10.12 Verify that the group of translations $T < GL(2, \mathbb{F}_p)$ (see Example 4.1(i)) is a normal subgroup of the Frobenius group $F_{(p-1)p}$ (see Exercise 4.4). Prove that the quotient group is isomorphic to $\mathbb{F}_p^\times$.

10.13 In $\mathbb{Q}/\mathbb{Z}$, what is the order of the coset of a/b, where $a, b \in \mathbb{Z}$, $b \neq 0$, and $(a, b) = 1$? Conclude that every element in $\mathbb{Q}/\mathbb{Z}$ has finite order, and that there are elements of arbitrarily large order.

10.14 Are there elements of infinite order in $\mathbb{R}/\mathbb{Z} \cong S$ (cf. Exercise 6.2)?

10.15 Let H be the Heisenberg group (see Exercise 8.15). Show that

$$H/Z(H) \cong \mathbb{R}^2 \ .$$

10.16 • In Example 10.6, let

$$H = \langle \{(1\ 4)(2\ 5)(3\ 6),\ (1\ 5\ 2\ 4)(3\ 6)\} \rangle \ .$$

Prove directly that $H \cong D_4$. Show that the composition

$$H \xrightarrow{i} G_{72} \xrightarrow{p} L \ ,$$

where i is the inclusion map and p is the canonical homomorphism, is an isomorphism. Is G_{72} isomorphic to $H \times K$?

10.17 • In $SL(2, \mathbb{C})$, let

$$\alpha = \begin{pmatrix} e^{2\pi i/3} & 0 \\ 0 & e^{4\pi i/3} \end{pmatrix}, \qquad \beta = \begin{pmatrix} 0 & 1 \\ -1 & 0 \end{pmatrix} \ .$$

Verify that $\beta\alpha\beta^{-1} = \alpha^{-1}$ and that $G_{12} := \langle \alpha, \beta \rangle$ has order 12.

10.18 Let G be a group such that $G/Z(G)$ is cyclic. Prove that G is abelian.

10.19 Show that $f_3 : PGL(2, \mathbb{F}_3) \to S_4$ is an isomorphism.

10.20 What is the order of $PSL(2, \mathbb{F}_p)$?

10.21 • Let F be a finite field. How many squares are there in F, that is, elements of the form a^2, $a \in F$? Suggestion: use Exercise 6.19.

10.22 (a) Suppose that H is a normal subgroup of a group G. Show that if $\alpha \in H$ then H contains the entire conjugacy class of α.

(b) • In Example 9.1(ii), we determined the conjugacy classes of A_5. Use this computation to prove that A_5 has no normal subgroups other than $\{1\}$ and A_5 itself.

10.23 • Let G be a group, $H \subset G$, a subgroup, and set $X = G/H$. Then G acts on X by multiplication on the left. As discussed in Chapter 8, this determines a homomorphism $\sigma : G \to S_X$, which is given by

$$\sigma(\alpha) \cdot \beta H = \alpha \beta H .$$

(a) Show that $\ker \sigma \subset H$;

(b) If $K \subset H$ is a normal subgroup of G, prove that $K \subset \ker \sigma$.

10.24 Let

$$\alpha = \begin{pmatrix} 1 & 1 \\ 0 & 1 \end{pmatrix}, \quad \beta = \begin{pmatrix} 0 & 1 \\ -1 & 0 \end{pmatrix}, \quad \gamma = \begin{pmatrix} 3 & 1 \\ 2 & 1 \end{pmatrix}$$

in $G = SL(2, \mathbb{F}_5)$. Then $g = \{\alpha, \beta\}$ generates G. Verify that $h = \{\beta, \gamma\}$ generates a subgroup H of index 5 (see Exercise 4.11). The group G acts on G/H. The five cosets of H in G each contain one of the powers of α. So label the cosets 1, 2, 3, 4, 5 by letting coset i be the one containing

$$\alpha^i = \begin{pmatrix} 1 & i \\ 0 & 1 \end{pmatrix} .$$

This defines an action of G on $\{1, 2, 3, 4, 5\}$, in other words, a homomorphism of G into S_5.

(a) Prove that the image of G is A_5.

(b) Identify the kernel of the mapping and prove that it induces an isomorphism of $PSL(2, \mathbb{F}_5)$ with A_5.

10.25 Let G be a group, and H and K normal subgroups. Suppose that $K < H$.

(a) Verify that H/K is a normal subgroup of G/K.

(b) (Second Isomorphism Theorem) Prove that $(G/K)\big/(H/K) \cong G/H$.

Chapter 11

Sylow Subgroups

The Sylow Theorems

The Sylow subgroups of a finite group G are a class of subgroups that provide the first clues for discovering the structure of G. We shall see later in this chapter that the results we obtain about Sylow subgroups are enough to classify groups of small order.

DEFINITION 11.1   *Let G be a group of order ap^r, where $(a, p) = 1$. A p-subgroup of order p^r is called a Sylow p-subgroup of G.*

In other words, a Sylow p-subgroup is a p-subgroup of the maximal possible order. For example, $|S_4| = 24 = 2^3 \cdot 3$. So a Sylow 2-subgroup is one of order 8, such as D_4. A Sylow 3-subgroup is one of order 3, for example a cyclic subgroup generated by a 3-cycle. If we consider the permutation group G_{72} in Example 10.5, we have that $72 = 2^3 \cdot 3^2$. The subgroup $H \cong D_4$ in Exercise 10.16 is a Sylow 2-subgroup and the subgroup $K \cong (\mathbb{Z}/3\mathbb{Z})^2$, a Sylow 3-subgroup.

Our first result tells us that for each prime p that divides $|G|$ there exists a Sylow subgroup. It is a partial converse to Lagrange's Theorem. In order to prove it we need an arithmetic lemma.

LEMMA 11.1
Suppose $n = ap^r$ where $(a, p) = 1$. Then

$$\binom{n}{p^r} \equiv a \pmod{p}.$$

In particular, p does not divide $\binom{n}{p^r}$.

PROOF   As we saw in Exercise 1.4,

$$(1 + x)^p \equiv 1 + x^p \pmod{p}.$$

Arguing by induction,

$$(1 + x)^{p^r} \equiv 1 + x^{p^r} \pmod{p} .$$

Therefore

$$(1 + x)^n = \left((1 + x)^{p^r}\right)^a \equiv (1 + x^{p^r})^a = 1 + ax^{p^r} + \cdots + x^n \pmod{p} .$$

But the coefficient of x^{p^r} in the expansion of $(1 + x)^n$ is $\binom{n}{p^r}$. Therefore

$$\binom{n}{p^r} \equiv a \pmod{p} . \quad \blacksquare$$

THEOREM 11.1
Let G be a finite group. For each prime p dividing $|G|$ there exists a Sylow p-subgroup.

PROOF   Suppose $n := |G| = ap^r$, where $(a, p) = 1$. Let X be the set of all subsets of G with p^r elements. We know that

$$|X| = \binom{n}{p^r} .$$

Now G acts on X by

$$(\alpha, T) \mapsto \alpha T ,$$

where $\alpha \in G$ and $T \in X$. X decomposes into a disjoint union of orbits of G. According to the lemma, p does not divide $|X|$. So from Formula 9.2 we see that the order of at least one of these orbits is not divisible by p. Suppose O_T is such an orbit, and G_T the stabilizer of T. By Formula 9.1,

$$|G| = |O_T||G_T| .$$

Since $|G|$ is divisible by p^r, it follows that $|G_T|$ is divisible by p^r. But for any $\tau \in T$, we have $G_T \tau \subset T$, so that

$$|G_T| = |G_T \tau| \leq |T| = p^r .$$

Therefore

$$|G_T| = p^r$$

and G_T is a Sylow p-subgroup. $\blacksquare$

REMARK 11.1   The proof shows that in fact

$$G_T \tau = T ,$$

in other words T is a right coset of the Sylow p-subgroup G_T. It also shows that

$$|O_T| = a .$$

On the other hand, suppose H is any Sylow p-subgroup, and set

$$T = H\tau ,$$

for some $\tau \in G$. Then it is easy to see that $G_T = H$, and therefore $|O_T| = |G|/|H| = a$. So orbits O_T for such T, are precisely the orbits whose length is prime to p. ∎

Our second result says something about the number of Sylow p-subgroups.

THEOREM 11.2
Let n_p be the number of Sylow p-subgroups of G. Then

$$n_p \equiv 1 \pmod{p} .$$

PROOF   As pointed out in the remark above, each orbit of length a consists of right cosets of Sylow p-subgroups. Cosets of different subgroups are distinct: for if

$$H\sigma = K\tau , \quad \sigma, \tau \in G ,$$

then

$$H\sigma\tau^{-1} = K .$$

In particular, $1 = \beta(\sigma\tau^{-1})$ for some $\beta \in H$. It follows that $\sigma\tau^{-1} = \beta^{-1} \in H$, so that $H = K$.

We can now count the number of elements of X in such orbits: there are n_p Sylow p-subgroups and each one has a cosets. Therefore, the total number of elements of X in these orbits is an_p. Since the length of any other orbit is divisible by p, Formula 9.2 shows that

$$|X| \equiv an_p \pmod{p} .$$

But according to Lemma 11.1,

$$|X| \equiv a \pmod{p} .$$

Therefore since $(a, p) = 1$,

$$n_p \equiv 1 \pmod{p} . \quad ∎$$

If H is a Sylow p-subgroup of G, then so is every conjugate of H. In the example $G = S_4$, we saw that the Sylow 3-subgroups are cyclic. They are in fact all conjugate to each other, because all 3-cycles are conjugate in S_4. The third result tells us that this is not a coincidence.

THEOREM 11.3
Let G be a finite group. Then its Sylow p-subgroups are conjugate to one another.

PROOF   Let H be a Sylow p-subgroup of G. Each left coset of H has p^r elements. Thus the set of left cosets of H, $G/H \subset X$. Let K be another Sylow p-subgroup. We can look at the action of K (by left multiplication) on X and in particular on G/H. Then G/H decomposes into disjoint K-orbits. Since p does not divide $|G/H|$, Formula 9.2 again says that there must be an orbit whose order is not divisible by p. Suppose the coset αH, $\alpha \in G$, belongs to such an orbit. By Formula 9.1, the order of this orbit divides $|K| = p^r$. But this order is not divisible by p. So it must be 1, in other words,

$$K\alpha H = \alpha H .$$

This means that for any $\kappa \in K$,

$$\kappa\alpha \cdot 1 \in \alpha H ,$$

or equivalently

$$\alpha^{-1}\kappa\alpha \in H .$$

Therefore $\alpha^{-1}K\alpha \subset H$, and since $|K| = |H|$, in fact

$$\alpha^{-1}K\alpha = H .$$

Thus any two Sylow p-subgroups are conjugate.   ∎

This result suggests another way of counting the number of Sylow p-subgroups. If we let G act on X by conjugation, then the orbit of a Sylow p-subgroup H is the set of its conjugates. The stabilizer is the subgroup

$$N_G(H) := \left\{\alpha \in G \mid \alpha H\alpha^{-1} = H\right\} .$$

$N_G(H)$ is called the *normalizer* of H in G. Clearly $H \triangleleft N_G(H)$ and if $H \triangleleft G$, then $N_G(H) = G$. Now we can apply Formula 9.1 again to see that the number of conjugates of H is $|G|/|N_G(H)|$. Since

$$\frac{|G|}{|N_G(H)|} \cdot \frac{|N_G(H)|}{|H|} = \frac{|G|}{|H|} = a ,$$

it follows that the number of conjugates of H divides a. Therefore:

COROLLARY 11.1
n_p *divides a.*

Examples 11.1

(i) As we remarked, D_4 is a Sylow 2-subgroup of S_4:

$$D_4 = \{(1),\ (1\,2)(3\,4),\ (1\,3)(2\,4),\ (1\,4)(2\,3),\ (1\,2\,3\,4),\ (1\,4\,3\,2),\ (1\,3),\ (2\,4)\}.$$

The subgroup

$$V = \{(1),\ (1\,2)(3\,4),\ (1\,3)(2\,4),\ (1\,4)(2\,3)\}$$

is a normal subgroup of S_4. So every conjugate of D_4 contains V. Now $\{(1\,2)(3\,4),\ (1\,2\,3\,4)\}$ generates D_4. There are six 4-cycles, all conjugate to each other. A pair of them, inverse to each other, occur in D_4 and each of its conjugates. Therefore there are three subgroups of S_4 conjugate to D_4. This fits with Theorem 11.2 and Corollary 11.1:

$$3 \equiv 1 \quad (\mathrm{mod}\ 2) \qquad \text{and} \qquad 3 = [S_4 : D_4] = a\ .$$

(ii) In the group G_{72}, K is normal. So it is the only Sylow 3-subgroup. From Theorem 11.2 and Corollary 11.1, we know that

$$n_2 \equiv 1 \quad (\mathrm{mod}\ 2) \qquad \text{and} \qquad n_2 \,|\, 9\ .$$

So $n_2 = 1, 3$, or 9. The Sylow 2-subgroup

$$\begin{aligned} H =\{&(1),\ (4\,5),\ (1\,2),\ (1\,2)(4\,5),\ (1\,4)(2\,5)(3\,6),\\ &(1\,4\,2\,5)(3\,6),\ (1\,5\,2\,4)(3\,6),\ (1\,5)(2\,4)(3\,6)\}\end{aligned}$$

is isomorphic to D_4, with generators $\{(1\,5\,2\,4)(3\,6),\ (1\,4)(2\,5)(3\,6)\}$. Using *Mathematica* (see Chapter 8), or otherwise, we see that $(1\,5\,2\,4)(3\,6)$ has 18 conjugates in G_{72}, occurring in mutually inverse pairs. Therefore H has at least 9 conjugates. So $n_2 = 9$.   ☐

In the last section we shall see how to find a Sylow 2-subgroup of S_8.

Groups of Small Order

THEOREM 11.4
A group of order p^2, where p is prime, is cyclic or is isomorphic to $\mathbb{Z}/p\mathbb{Z} \times \mathbb{Z}/p\mathbb{Z}$.

PROOF   Let G be a group of order p^2 and assume that G is not cyclic. According to Theorem 9.2, the centre of G has order at least p. So take an element $\alpha \in Z(G)$,

$\alpha \neq 1$. Then $|\alpha| = p$. Now pick an element $\beta \notin \langle \alpha \rangle$. Since $|\beta| \mid p^2$ and G is not cyclic, it follows that $|\beta| = p$ too. And because $\alpha \in Z(G)$, $\alpha\beta = \beta\alpha$. Therefore the mapping

$$\mathbb{Z}/p\mathbb{Z} \times \mathbb{Z}/p\mathbb{Z} \to G$$

given by

$$(a, b) \mapsto \alpha^a \beta^b \ , \ a, b \in \mathbb{Z}/p\mathbb{Z}$$

is a well-defined homomorphism. This homomorphism is injective. But both groups have order p^2, so it is in fact an isomorphism. Thus G is either cyclic or isomorphic to $(\mathbb{Z}/p\mathbb{Z})^2$. ∎

For example a group of order 9 is either cyclic or isomorphic to $(\mathbb{Z}/3\mathbb{Z})^2$. The next result deals with groups of order $2p$.

THEOREM 11.5
A group of order $2p$, where $p \geq 3$ is prime, is isomorphic to $\mathbb{Z}/2p\mathbb{Z}$ or to D_p.

PROOF   Let G be a group of order $2p$. Its Sylow p-subgroup has order p and is therefore cyclic (see Corollary 10.3). Similarly, the Sylow 2-subgroup is cyclic of order 2. So let α be an element of G of order p, and β of order 2. Since $\langle \alpha \rangle$ is of index 2 in G, it is a normal subgroup by Remark 10.1. Therefore

$$\beta\alpha\beta = \alpha^k \ ,$$

for some $k, 0 < k < p$. Conjugating again with β, we get

$$\alpha = \beta^2 \alpha \beta^2 = \beta\alpha^k\beta = (\alpha^k)^k = \alpha^{k^2} \ .$$

Therefore $k^2 \equiv 1 \pmod{p}$, which means that $k = \pm 1$.
    Thus there are two cases: first,

$$\beta\alpha\beta = \alpha \ ,$$

which says that α and β commute. But then $\alpha\beta$ has order $2p$ (see Exercise 5.7) and G is cyclic.
    In the second case

$$\beta\alpha\beta = \alpha^{-1} \ .$$

This is the defining relation for D_p (Equation (7.1)). So G is isomorphic to D_p. ∎

In particular, for $p = 3$ this says again that a group of order 6 is either cyclic or isomorphic to S_3. And for $p = 5$, it says that a group of order 10 is either cyclic or isomorphic to D_5.
    It is time to sort out the groups of order 8.

THEOREM 11.6
Let G be a group of order 8. *Then G is isomorphic to* D_4, Q, $\mathbb{Z}/8\mathbb{Z}$, $\mathbb{Z}/4\mathbb{Z} \times \mathbb{Z}/2\mathbb{Z}$
or $(\mathbb{Z}/2\mathbb{Z})^3$.

PROOF   Suppose that G is not cyclic. Then its nontrivial elements have order 2
or 4. It is not hard to see that if all of these have order 2, then

$$G \cong (\mathbb{Z}/2\mathbb{Z})^3 .$$

So let $\alpha \in G$ be an element of order 4. Since the index of $\langle \alpha \rangle$ is 2, $\langle \alpha \rangle$ is a normal
subgroup. Pick an element $\beta \notin \langle \alpha \rangle$. Then

$$\beta \alpha \beta^{-1} = \alpha^k ,$$

where $k = \pm 1$. If $k = 1$, then α and β commute and G is abelian. There are two
possibilities: either $|\beta| = 2$ or $|\beta| = 4$. In the first case, arguing as in the proof of
Theorem 11.4, we see that

$$G \cong \mathbb{Z}/4\mathbb{Z} \times \mathbb{Z}/2\mathbb{Z} .$$

In the second case, we must have that $\beta^2 = \alpha^2$. Therefore $|\alpha\beta| = 2$. So replacing β
by $\alpha\beta$ we are back to the first case.
    Now suppose that

$$\beta \alpha \beta^{-1} = \alpha^{-1} .$$

Thus G is not abelian. If $|\beta| = 2$, then the relation tells us that $G \cong D_4$ (see
Equation (7.1)). This leaves us with the case $|\beta| = 4$. According to Theorem 9.2,
$Z(G)$ is not trivial. It is not hard to see that

$$|Z(G)| = 2 ,$$

and that if $\gamma \in G$ has order 4, then

$$Z(G) \subset \langle \gamma \rangle .$$

So let ϵ generate $Z(G)$. Then $\alpha^2 = \epsilon$, $\beta^2 = \epsilon$, and therefore

$$(\beta\alpha)^2 = (\beta\alpha)(\alpha^{-1}\beta) = \beta^2 = \epsilon .$$

It follows that $|\beta\alpha| = 4$. Furthermore,

$$\beta\alpha = \alpha^{-1}\beta = \alpha^3\beta = \epsilon(\alpha\beta) .$$

Comparing this with the description of Q in Exercise 4.5, we see that $G \cong Q$.   ∎

    The following result is the key to classifying groups of order 12.

THEOREM 11.7

Let G be a group of order 12. *Then G has a normal subgroup of order* 3 *or* $G \cong A_4$.

PROOF   Write $12 = 2^2 \cdot 3$. According to Theorem 11.2, $n_3 \equiv 1$ (mod 3). According to Corollary 11.1, n_3 divides 4. So $n_3 = 1$ or 4. Suppose that $n_3 \neq 1$, in other words that G does not have a normal subgroup of order 3. Let H be one of the subgroups of order 3. Then G acts on the set of four left cosets, G/H, by multiplication on the left. This defines a homomorphism

$$\sigma : G \rightarrow S_4 .$$

What is the kernel of σ? Well, $\sigma(\alpha) = 1$ means that for all $\beta \in G$,

$$\alpha(\beta H) = \beta H .$$

Equivalently

$$\beta^{-1}\alpha\beta H = H$$

or

$$\beta^{-1}\alpha\beta \in H$$

for all $\beta \in G$. But then

$$\alpha \in \beta H \beta^{-1}$$

for all $\beta \in G$. By assumption, H has four distinct conjugates, and the intersection of two distinct subgroups of order 3 is trivial. Therefore $\alpha = 1$ and σ is injective. So G is isomorphic to a subgroup of S_4 of order 12, and the only such subgroup is A_4.
∎

With this result it is not hard to classify groups of order 12 (see Exercise 11.15). It turns out that up to isomorphism there are five groups: $\mathbb{Z}/12\mathbb{Z}$, $\mathbb{Z}/3\mathbb{Z} \times V$, D_6, A_4, and G_{12} (see Exercise 4.17).

Example 11.2

Let G be a group of order $15 = 3 \cdot 5$. Then $n_3 \equiv 1$ (mod 3) and $n_3 \mid 5$. It follows that $n_3 = 1$. Similarly, $n_5 \equiv 1$ (mod 5) and $n_5 \mid 3$. So $n_5 = 1$ too. Thus G has only one Sylow 3-subgroup and only one Sylow 5-subgroup and both are normal. Let α be an element of order 3 and β of order 5. Applying Exercise 10.8, we see that

$$\alpha\beta = \beta\alpha .$$

But then by Exercise 5.7,

$$|\alpha\beta| = |\alpha||\beta| = 15 .$$

So G is cyclic. Thus every group of order 15 is cyclic.   ∎

A List

We can now begin a list, up to isomorphism, of all the groups of very small order.

* order 1: $\{1\}$.

* order 2: Since 2 is prime, all groups of order 2 are cyclic and therefore isomorphic to $\mathbb{Z}/2\mathbb{Z}$.

* order 3: Just as for 2, all groups of order 3 are isomorphic to $\mathbb{Z}/3\mathbb{Z}$.

* order 4: It is easy to see that a group of order 4 is cyclic or isomorphic to V. Notice that both are abelian.

* order 5: $\mathbb{Z}/5\mathbb{Z}$.

* order 6: By Theorem 11.5, there are two groups of order 6: $\mathbb{Z}/6\mathbb{Z} \cong \mathbb{Z}/2\mathbb{Z} \times \mathbb{Z}/3\mathbb{Z}$ and S_3. S_3 is the smallest nonabelian group.

* order 7: $\mathbb{Z}/7\mathbb{Z}$.

* order 8: By Theorem 11.6, there are five groups of order 8: D_4, Q, $\mathbb{Z}/8\mathbb{Z}$, $\mathbb{Z}/4\mathbb{Z} \times \mathbb{Z}/2\mathbb{Z}$, and $(\mathbb{Z}/2\mathbb{Z})^3$.

* order 9: From Theorem 11.4, we know that every group of order 9 is either cyclic or isomorphic to $(\mathbb{Z}/3\mathbb{Z})^2$.

* order 10: By Theorem 11.5, any group of order 10 is isomorphic to $\mathbb{Z}/10\mathbb{Z}$ or D_5.

* order 11: $\mathbb{Z}/11\mathbb{Z}$.

* order 12: As mentioned following Theorem 11.7, the groups of order 12 are $\mathbb{Z}/12\mathbb{Z}$, $\mathbb{Z}/3\mathbb{Z} \times V$, D_6, A_4, and G_{12}.

* order 13: $\mathbb{Z}/13\mathbb{Z}$.

* order 14: Again Theorem 11.5 shows that the groups of order 14 are $\mathbb{Z}/14\mathbb{Z}$ and D_7.

* order 15: Example 11.2 shows that all groups of order 15 are cyclic.

Notice that there is no nonabelian group of odd order in this list. Can you find a nonabelian group of least odd order?

A Calculation

Let us look at the Sylow subgroups of S_8. First factor 8!.

```
In[1]:= FactorInteger[8!]
```

$$Out[1] = \{\{2,7\},\{3,2\},\{5,1\},\{7,1\}\}$$

In other words,

$$8! = 2^7 3^2 5 7.$$

The Sylow 5-subgroups and Sylow 7-subgroups are cyclic, generated by 5-cycles and 7-cycles, respectively. A Sylow 3-subgroup is generated by two disjoint 3-cycles. A Sylow 2-subgroup has order $2^7 = 128$ and is harder to find. The order of any element in it is a power of 2. So let us begin our list of generators with an 8-cycle, say

$$(1\,2\,3\,4\,5\,6\,7\,8).$$

If we add an arbitrary 4-cycle, we will get a group that is too big. Now the square of this 8-cycle is

$$(1\,3\,5\,7)(2\,4\,6\,8).$$

So let us take $(1\,3\,5\,7)$ as the second generator.

```
In[2]:= H = Group[ P[{1,2,3,4,5,6,7,8}],
                   P[{1,3,5,7}] ]
```

$$Out[2] = \langle\,(1\ 2\ 3\ 4\ 5\ 6\ 7\ 8),\ (1\ 3\ 5\ 7)\,\rangle$$

```
In[3]:= Order[H]
```

$$Out[3] = 32$$

So we have to add another element to our list of generators. If we take an arbitrary 2-cycle, we will get the whole group S_8. But the square of our 4-cycle is

$$(1\,5)(3\,7).$$

So let us add $(1\,5)$ to our set of generators.

```
In[4]:= H = Group[ P[{1,2,3,4,5,6,7,8}], P[{1,3,5,7}],
                   P[{1,5}] ]
```

$$Out[4] = \langle\,(1\ 2\ 3\ 4\ 5\ 6\ 7\ 8),\ (1\ 3\ 5\ 7),\ (1\ 5)\,\rangle$$

In [5] := Order [H]

Out [5] = 128

We have found a Sylow 2-subgroup!

Theorem 11.2 and Corollary 11.1 tell us that the number of Sylow 2-subgroups is odd and divides $8!/128 = 315$. It would be helpful to know how many there really are. By the second Sylow Theorem (Theorem 11.3), all Sylow 2-subgroups are conjugate to one another. So we must determine the number of subgroups of S_8 conjugate to our subgroup H. The function NumberOfConjugates will do this. An 8-cycle and a transposition generate S_8:

In [6] := S8 = Group[P[{1,2,3,4,5,6,7,8}], P[{1,2}]]

Out [6] = ⟨ (1 2 3 4 5 6 7 8), (1 2) ⟩

Then

In [7] := NumberOfConjugates [S8, H]

Out [7] = 315

Exercises

11.1 Write down the Sylow 2-subgroups of S_4. Show directly that they are conjugate to each other.

11.2 Find a Sylow p-subgroup of S_6 for each prime p dividing 6!.

11.3 Verify that the group of translations T (see Example 4.1(i)) is a Sylow p-subgroup of $GL(2, \mathbb{F}_p)$. Find another Sylow p-subgroup. What is n_p?

11.4 Let $p > 2$ be a prime number. What is the order of a Sylow p-subgroup of S_{2p}? Give an example of such a subgroup by giving a set of generators for it.

11.5 With the notation and assumptions of the proof of the first Sylow Theorem (Theorem 11.1), let H be a Sylow p-subgroup of G. What is the stabilizer of $\alpha H \tau$? Write $\alpha H \tau$ as a right coset of a Sylow p-subgroup.

11.6 Suppose that G is a group of order pq, where p and q are prime, $p < q$ and $p \nmid (q - 1)$. Prove that G is cyclic.

11.7 Determine all numbers $n < 70$ that are the product of two primes satisfying the conditions of the previous exercise.

11.8 • Let G be a p-group. Show that G has a subgroup of every order that divides $|G|$.

11.9 • Let G be a finite group, and p a prime dividing $|G|$. Show that G has an element of order p.

11.10 Prove that a group of order 4 is either cyclic or isomorphic to V.

11.11 Classify all groups of order 26.

11.12 Classify all groups of order 21.

11.13 Let G be a group of order 8. Suppose every element except 1 has order 2. Prove that G is abelian and

$$G \cong (\mathbb{Z}/2\mathbb{Z})^2 .$$

11.14 Let G be a group of order 8.

(a) Suppose that $|Z(G)| \geq 4$. Show that G is abelian.

(b) Suppose that G is not abelian, and that $\alpha \in G$ has order 4. Prove that

$$Z(G) \subset \langle \alpha \rangle .$$

11.15 Suppose that G is a group of order 12 and $G \not\cong A_4$. By Theorem 11.7, G has a normal subgroup H of order 3. Let K be a Sylow 2-subgroup of G. Then K acts on H by conjugation.

(a) Show that the kernel of this action has order 2 if G is not abelian.

(b) Prove that $G \cong D_6$ or $G \cong \mathbb{Z}/3\mathbb{Z} \times V$.

(c) Suppose that K is cyclic. Prove that G is cyclic or that $G \cong G_{12}$.

11.16 How many Sylow 2-subgroups does S_5 have? S_6?

11.17 Prove that the construction in the previous section gives 315 Sylow 2-subgroups of S_8.

11.18 How large is the centre of a Sylow 2-subgroup of S_8?

11.19 What is the order of a Sylow p-subgroup of S_{p^2}, for p prime? Give an example of one.

Chapter 12

Simple Groups

Composition Series

If we were to continue classifying groups of small order, we would get a table like this one.

Order of Groups	Number of Groups
16	14
17	1
18	5
19	1
20	5
21	2
22	2
23	1
24	15

As the table suggests, when the order has many prime factors, there tend to be many groups of that order. So the order of a group does not tell you very much about it. Classifying groups in this way is not very enlightening. A better way to understand groups is to analyze how they are built up out of certain "building blocks." The building blocks are called simple groups.

DEFINITION 12.1   *A group G is simple if it has no normal subgroups other than {1} and G itself.*

We have seen some simple groups already: the groups of prime order. By Lagrange's Theorem, they have no nontrivial subgroups at all. To understand how a group is built out of simple groups, we need the following observation.

THEOREM 12.1

Let G be a group and $K \lhd G$. Let $p : G \to \bar{G} := G/K$ be the canonical homomorphism. Then there is a 1-to-1 correspondence between subgroups of G containing K and subgroups of $\bar{G}$ given by

$$H \mapsto p(H) = H/K \,,$$

where $K < H < G$. Furthermore H is normal in G if and only if $p(H)$ is normal in $\bar{G}$.

PROOF   For any subgroup $\bar{H} < \bar{G}$ the set

$$H := p^{-1}(\bar{H}) := \{\alpha \in G \mid p(\alpha) \in \bar{H}\}$$

is a subgroup of G. For $1 \in H$, so $H \neq \emptyset$. If $\alpha \in H$, then $p(\alpha) \in \bar{H}$, which means that $p(\alpha^{-1}) = p(\alpha)^{-1} \in \bar{H}$ so that $\alpha^{-1} \in H$. And if $\alpha, \beta \in H$, then $p(\alpha\beta) = p(\alpha)p(\beta) \in \bar{H}$ so that $\alpha\beta \in H$.

Now given $\bar{H} < \bar{G}$, $p^{-1}(\bar{H}) > K$ and $p(p^{-1}(\bar{H})) = \bar{H}$. And if $K < H < G$, then $p^{-1}(p(H)) = H$. Thus the correspondence

$$H \leftrightarrow p(H)$$

is 1-to-1. Suppose $H \lhd G$. Given $\bar{\alpha} \in \bar{G}$, pick $\alpha \in G$ with $p(\alpha) = \bar{\alpha}$. Then

$$p(H) = p(\alpha H \alpha^{-1}) = p(\alpha)p(H)p(\alpha^{-1}) = \bar{\alpha} p(H) \bar{\alpha}^{-1} \,,$$

so that $p(H)$ is normal in $\bar{G}$. Conversely, if $\bar{H} \lhd \bar{G}$ and $\alpha \in G$, then

$$p(\alpha \, p^{-1}(\bar{H}) \alpha^{-1}) = p(\alpha) \bar{H} p(\alpha)^{-1} = \bar{H} \,.$$

Therefore $\alpha \, p^{-1}(\bar{H}) \alpha^{-1} = p^{-1}(\bar{H})$ and $p^{-1}(\bar{H}) \lhd G$. ∎

Now we can see how a finite group G is built up from simple groups. Let H_1 be a nontrivial proper normal subgroup that is as large as possible. In other words, there should be no proper normal subgroup containing H_1. Then G/H_1 has no nontrivial proper normal subgroups and is therefore simple. Next, pick a nontrivial proper normal subgroup H_2 of H_1 that is as large as possible. Again, by the theorem above, H_1/H_2 is simple. Continue in this way until you have a nontrivial subgroup H_{n-1}, which is simple itself. This will happen because the orders of the subgroups H_1, H_2, ... are strictly decreasing. This descending sequence of subgroups describes how G is built up out of simple groups.

DEFINITION 12.2   *A composition series for a group G is a sequence of subgroups*

$$G = H_0 \rhd H_1 \rhd \cdots \rhd H_{n-1} \rhd H_n = \{1\}$$

where H_i/H_{i+1} is a simple group for $0 \le i < n$. The quotient groups H_i/H_{i+1} are called composition factors of G.

For example, let us write down a composition series for S_4. We have $A_4 \lhd S_4$ and $V \lhd A_4$. Since $V \cong \mathbb{Z}/2\mathbb{Z} \times \mathbb{Z}/2\mathbb{Z}$, we have one more term, $\mathbb{Z}/2\mathbb{Z} \lhd V$. Thus our composition series is

$$S_4 \rhd A_4 \rhd V \rhd \mathbb{Z}/2\mathbb{Z} \rhd \{1\} \, . \qquad (12.1)$$

The composition factors are

$$S_4/A_4 \cong \mathbb{Z}/2\mathbb{Z} \qquad A_4/V \cong \mathbb{Z}/3\mathbb{Z} \qquad V\big/(\mathbb{Z}/2\mathbb{Z}) \cong \mathbb{Z}/2\mathbb{Z} \qquad \mathbb{Z}/2\mathbb{Z} \, .$$

A composition series for S_5 is

$$S_5 \rhd A_5 \rhd \{1\} \, .$$

That is all, because we saw in Exercise 10.22 that A_5 is a simple group.

THEOREM 12.2 (Jordan–Hölder)
Let G be a finite group. Then G has a composition series. The composition factors are unique in the following sense. If

$$G = H_0 \rhd H_1 \rhd \cdots \rhd H_{m-1} \rhd H_m = \{1\}$$

and

$$G = K_0 \rhd K_1 \rhd \cdots \rhd K_{n-1} \rhd K_n = \{1\}$$

are two composition series for G, then $m = n$ and the composition factors are the same up to permutation, that is there is a permutation σ of $\{1, 2, \ldots, n\}$ such that

$$H_i/H_{i+1} \cong K_{\sigma(i)}/K_{\sigma(i+1)} \, ,$$

for $0 \le i < n$.

We have seen that composition series exist, but shall not prove that they are unique. Clearly it becomes important to know what the simple groups are. We shall show that two families of groups are simple: A_n, for $n \ge 5$, and $PSL(2, \mathbb{F}_p)$, for p prime, $p > 2$. That A_n is simple will be used in Chapter 20 to prove that in general a polynomial equation of degree 5 or more cannot be solved by radicals.

Simplicity of A_n

We know that A_3, which is cyclic of order 3, is simple and that A_4 is not. In Exercise 10.22 we saw that A_5 is simple. In fact, for all $n \ge 5$, A_n is simple. First we show that A_n is generated by the set of all 3-cycles (cf. Exercise 3.12).

THEOREM 12.3
The set of 3-cycles generates A_n, $n \geq 3$.

PROOF   By definition, an even permutation can be written as a product of an even number of transpositions. So it is sufficient to show that a product α of two transpositions is a product of 3-cycles. Now there are two possibilities for α. Either the two transpositions have a symbol in common or they do not:

$$\alpha = (i\ j)(j\ k)$$

or

$$\alpha = (i\ j)(k\ l)\ .$$

Now

$$(i\ j)(j\ k) = (i\ j\ k)\ ,$$

and

$$(i\ j)(k\ l) = (i\ j)(j\ k)(j\ k)(k\ l) = (i\ j\ k)(j\ k\ l)\ .$$

So α is indeed a product of 3-cycles and therefore any even permutation is a product of 3-cycles. ∎

Suppose that $N \lhd A_n$. What we shall do is to prove that N must contain all the 3-cycles and therefore be A_n. The following lemma tells us that if one 3-cycle belongs to N, then all of them do.

LEMMA 12.1
Suppose that $N \lhd A_n$, $n \geq 5$, which contains one 3-cycle. Then N contains all 3-cycles. Equivalently, the 3-cycles form a single conjugacy class in A_n.

PROOF   Suppose that the 3-cycle $(i\ j\ k) \in N$. Since $n \geq 5$, there exist $l, m \leq n$ different from i, j, k. Now let α be any other 3-cycle. As we saw in Theorem 8.1, there exists a permutation $\beta \in S_n$ such that

$$\beta \alpha \beta^{-1} = (i\ j\ k)\ .$$

If β is even, then we are done. Otherwise we can replace β by $(l\ m)\beta$, since

$$(l\ m)\beta\alpha\beta^{-1}(l\ m) = (l\ m)(i\ j\ k)(l\ m) = (i\ j\ k)\ . \quad ∎$$

So we want to show that N contains a 3-cycle. To do this we look at *commutators* $\gamma = \alpha\beta\alpha^{-1}\beta^{-1}$, where $\alpha \in N$ and $\beta \in A_n$ (see also Exercise 13.12). Since N is normal, $\beta\alpha^{-1}\beta^{-1} \in N$, and therefore $\gamma \in N$.

THEOREM 12.4
The alternating groups A_n, for $n \neq 4$ are simple.

PROOF   Since A_3 is cyclic of order 3, it is simple. So we can assume that $n \geq 5$. We now proceed by induction on n. In Exercise 10.22 we saw that A_5 is simple. So let n be greater than 5 and assume that A_r is simple for all $r < n$. Let N be a nontrivial normal subgroup of A_n. A_n acts on $\{1, 2, \ldots, n\}$ and the stabilizer of any number is isomorphic to A_{n-1}. So for any s, the stabilizer N_s is isomorphic to a normal subgroup of A_{n-1}. If N_s is not trivial, then by the induction assumption it must be isomorphic to A_{n-1} itself. In particular it contains a 3-cycle. Therefore N contains a 3-cycle. Then Lemma 12.1 proves that N contains all 3-cycles. But according to Theorem 12.3, the set of 3-cycles generates A_n. Therefore $N = A_n$.

It remains to prove that for some s, N_s is not trivial. Suppose $\alpha \in N, \alpha \neq 1$. If $\beta \in A_n$ is a 3-cycle, then $\alpha \beta \alpha^{-1}$ is a 3-cycle too. Therefore the commutator $\gamma = \alpha \beta \alpha^{-1} \beta^{-1} \in N$ is a product of two 3-cycles. Suppose that

$$\gamma = (h \, i \, j)(k \, l \, m) \, .$$

If h, i, j, k, l, m are not distinct, then since $n \geq 6$, γ has a fixed point s, and thus N_s is not trivial. If they are distinct, let $\delta = (i \, j \, k)$. Then

$$\epsilon = \gamma \delta \gamma^{-1} \delta^{-1} = (i \, k \, h \, l \, j) \, .$$

Now $\epsilon \in N$ as well and it does have a fixed point, namely m. So N_m is not trivial.

We have shown therefore that A_n is simple and by the principle of induction, the theorem is proved.   ∎

Simplicity of $PSL(2, \mathbb{F}_p)$

Another family of groups that we can prove is simple is $PSL(2, \mathbb{F}_p)$, for p prime. In Chapter 4 we computed generators for $SL(2, \mathbb{F}_p)$. The first step is to refine this result.

THEOREM 12.5
$SL(2, \mathbb{F}_p)$ is generated by

$$\left\{ \begin{pmatrix} 1 & 1 \\ 0 & 1 \end{pmatrix}, \begin{pmatrix} 0 & 1 \\ -1 & 0 \end{pmatrix} \right\}$$

or equivalently, by

$$\left\{ \begin{pmatrix} 1 & 1 \\ 0 & 1 \end{pmatrix}, \begin{pmatrix} 1 & 0 \\ 1 & 1 \end{pmatrix} \right\} \, .$$

PROOF   We have the relations

$$\begin{pmatrix} 0 & -1 \\ 1 & 0 \end{pmatrix} \begin{pmatrix} 1 & 1 \\ 0 & 1 \end{pmatrix} \begin{pmatrix} 0 & 1 \\ -1 & 0 \end{pmatrix} = \begin{pmatrix} 1 & 0 \\ -1 & 1 \end{pmatrix} \tag{12.2}$$

$$\begin{pmatrix} 1 & 1 \\ 0 & 1 \end{pmatrix} \begin{pmatrix} 1 & 0 \\ -1 & 1 \end{pmatrix} \begin{pmatrix} 1 & 1 \\ 0 & 1 \end{pmatrix} = \begin{pmatrix} 0 & 1 \\ -1 & 0 \end{pmatrix} . \tag{12.3}$$

as in Chapter 4. These tell us that if one pair of matrices generates $SL(2, \mathbb{F}_p)$, then so does the other. According to Exercise 4.8, all we need to show is that we can write

$$\begin{pmatrix} c & 0 \\ 0 & c^{-1} \end{pmatrix} , \; c \in \mathbb{F}_p^\times ,$$

in terms of these matrices. Now for any $a, b \in \mathbb{F}_p$, we have

$$\begin{pmatrix} 1 & 0 \\ b & 1 \end{pmatrix} \begin{pmatrix} 1 & a \\ 0 & 1 \end{pmatrix} = \begin{pmatrix} 1 & a \\ b & 1 + ab \end{pmatrix} .$$

Taking $a = 1$, $b = c^{-1} - 1$ gives us the matrix

$$\begin{pmatrix} 1 & 1 \\ c^{-1} - 1 & c^{-1} \end{pmatrix} .$$

Taking $a = -c^{-1}$, $b = c - 1$ gives us the matrix

$$\begin{pmatrix} 1 & -c^{-1} \\ c - 1 & c^{-1} \end{pmatrix} .$$

If we multiply these two together we have

$$\begin{pmatrix} 1 & 1 \\ c^{-1} - 1 & c^{-1} \end{pmatrix} \begin{pmatrix} 1 & -c^{-1} \\ c - 1 & c^{-1} \end{pmatrix} = \begin{pmatrix} c & 0 \\ 0 & c^{-1} \end{pmatrix} .$$

Therefore the pair

$$\left\{ \begin{pmatrix} 1 & 1 \\ 0 & 1 \end{pmatrix} , \begin{pmatrix} 1 & 0 \\ 1 & 1 \end{pmatrix} \right\}$$

generates $SL(2, \mathbb{F}_p)$.   ∎

REMARK 12.1   For convenience, write $G := SL(2, \mathbb{F}_p)$. In Chapter 8 we saw that G acts on $P(\mathbb{F}_p)$ by fractional linear transformations. From the proof of Theorem 8.2 it follows that

$$H := G_0 \cap G_\infty = \left\{ \begin{pmatrix} c & 0 \\ 0 & c^{-1} \end{pmatrix} \Big| c \in \mathbb{F}_p^\times \right\} .$$

If we pick a $c \in \mathbb{F}_p^\times$ and set

$$\delta = \begin{pmatrix} c & 0 \\ 0 & c^{-1} \end{pmatrix}$$

then the corresponding fractional linear transformation is given by

$$s_\delta(x) = c^2 x . \quad \blacksquare$$

THEOREM 12.6
For $p > 3$, the groups $PSL(2, \mathbb{F}_p)$ are simple.

PROOF Let $\bar{N} \lhd PSL(2, \mathbb{F}_p)$ be a nontrivial subgroup and $N \lhd SL(2, \mathbb{F}_p)$ be its inverse image under $SL(2, \mathbb{F}_p) \to PSL(2, \mathbb{F}_p)$ (see Theorem 12.1). We want to show that $N = G$. To do this we shall show that

$$\tau := \begin{pmatrix} 1 & 1 \\ 0 & 1 \end{pmatrix} \in N . \tag{12.4}$$

Why will this do the trick? Well, since N is normal, Relation (12.2) above tells us that

$$\begin{pmatrix} 1 & 0 \\ -1 & 1 \end{pmatrix} \in N$$

and therefore by Theorem 12.5, $N = SL(2, \mathbb{F}_p)$.
   We will prove (12.4) by showing that

(i) N acts transitively on $P(\mathbb{F}_p)$;

(ii) N_∞ acts transitively on $P(\mathbb{F}_p) \setminus \{\infty\}$.

The first statement tells us that $p + 1$ divides $|N|$ and the second that p divides $|N_\infty|$ and therefore $|N|$ as well. So $p(p+1)$ divides $|N|$. Now $|G| = (p-1)p(p+1)$ (see Example 10.4(ii)). Therefore $|G/N|$ divides $p - 1$. In particular, $(|G/N|, p) = 1$. But $|\tau| = p$. So the order of $\bar{\tau}$ in G/N must divide p. This is only possible if $\bar{\tau} = \bar{1}$, in other words, $\tau \in N$.
   First we show that N is transitive on $P(\mathbb{F}_p)$. So given $c \in P(\mathbb{F}_p)$, we must find a $\gamma \in N$ such that

$$s_\gamma(0) = c .$$

Since $\bar{N} \neq \{1\}$, there exists $\alpha \in N$ with $s_\alpha \neq 1$. Thus for some $a \in P(\mathbb{F}_p)$,

$$b := s_\alpha(a) \neq a .$$

Since G is doubly transitive (see Exercise 8.24), there exists $\beta \in G$ with

$$s_\beta(a) = 0 , \quad s_\beta(b) = c .$$

If we set $\gamma = \beta\alpha\beta^{-1} \in N$, then

$$s_\gamma(0) = s_\beta s_\alpha s_\beta^{-1}(0) = c .$$

Thus N is transitive on $P(\mathbb{F}_p)$.

Secondly, we prove that N_∞ is transitive on $P(\mathbb{F}_p) \setminus \{\infty\}$. We will do this by showing that the orbit of 0 has length p. First let us check that N_∞ is not trivial. Take $\alpha \in N$ such that

$$s_\alpha(0) = \infty . \tag{12.5}$$

Then for any $\delta \in H = G_0 \cap G_\infty$,

$$s_{\delta\alpha\delta^{-1}}(0) = \infty .$$

Since $\delta\alpha\delta^{-1} \in N$, there is more than one element in N satisfying (12.5). Let β be a second such element. It follows that $\alpha\beta^{-1} \in N_\infty$ and $\alpha\beta^{-1} \neq 1$. Now pick a $\gamma \in N_\infty$, $\gamma \neq 1$. Then for

$$\delta = \begin{pmatrix} c & 0 \\ 0 & c^{-1} \end{pmatrix} , \quad c \in \mathbb{F}_p^\times,$$

we have

$$s_{\delta\gamma\delta^{-1}}(0) = c^2 s_\gamma(0) .$$

If necessary, by replacing γ with $\alpha\gamma\alpha^{-1}$ for a suitably chosen $\alpha \in G_\infty$, we can assure that $s_\gamma(0) \neq 0$. Now there are $(p-1)/2$ squares in $\mathbb{F}_p^\times$ (see Exercise 10.21). So O_0, the orbit of 0 under N_∞, must have length at least $(p-1)/2 + 1 = (p+1)/2$. We know that

$$|O_0| \,\big|\, |N_\infty| \,\big|\, |G_\infty| = (p-1)p .$$

Since $(p+1)/2$ does not divide $p - 1$, it follows that $|O_0| = p$. In other words, N_∞ is transitive on $P(\mathbb{F}_p) \setminus \{\infty\}$. ∎

Exercises

12.1 Write down the composition series for Q and for D_4, with their composition factors.

12.2 Find a composition series for G_{72}. What are the composition factors?

12.3 Find a composition series for a Sylow 2-subgroup of S_8.

12.4 Let G be a group of order pq, where p and q are distinct primes. Show that G is not simple.

12.5 Prove that the product of two 3-cycles is either

   (a) a product of two disjoint 3-cycles, or

   (b) a 5-cycle, or

(c) a product of two disjoint transpositions, or

(d) a 3-cycle.

12.6 Verify that if $\gamma = (h\,i\,j)(k\,l\,m)$, $\delta = (i\,j\,k) \in S_6$, where h, i, j, k, l, m are distinct, then

$$\gamma\delta\gamma^{-1}\delta^{-1} = (i\,k\,h\,l\,j)\,.$$

12.7 Suppose $\gamma \in A_n$, $\gamma \neq 1$, $n \geq 5$. Show that if γ has at least two fixed points, then there exists a 3-cycle δ such that their commutator is a 3-cycle.

12.8 Let $p > 2$ be prime.

(a) Set

$$\alpha = \begin{pmatrix} 0 & 1 \\ -1 & 0 \end{pmatrix}\,.$$

Show that $s_\alpha \in A_{p+1}$. You may use the result that the congruence $x^2 \equiv -1 \pmod{p}$ has a solution if and only if $p \equiv 1 \pmod 4$.

(b) Prove that $f_\alpha(SL(2, \mathbb{F}_p)) < A_{p+1}$.

12.9 A matrix

$$\alpha = \begin{pmatrix} a & b \\ c & d \end{pmatrix} \in SL(2, \mathbb{Z})$$

is congruent to the identity matrix I modulo p, if

$$a \equiv 1 \pmod{p} \qquad b \equiv 0 \pmod{p}$$
$$c \equiv 0 \pmod{p} \qquad d \equiv 1 \pmod{p}$$

Let

$$\Gamma(p) := \{\alpha \in SL(2, \mathbb{Z}) \mid \alpha \equiv I \pmod{p}\}\,.$$

(a) Show that $\Gamma(p) \lhd SL(2, \mathbb{Z})$.

(b) Prove that

$$SL(2, \mathbb{Z})/\Gamma(p) \cong SL(2, \mathbb{F}_p)\,.$$

12.10 Suppose that H is a normal subgroup of S_n, $n > 4$. Prove that $H = S_n$ or $H = A_n$ or H is trivial.

12.11 Can you generalize Theorem 12.6 to any finite field F, with $|F| > 3$?

12.12 Suppose that G is a finite group and H a proper subgroup such that $|G| \nmid |G/H|!$. Prove that H contains a nontrivial normal subgroup of G. Suggestion: use Exercise 10.23.

12.13 Let G be a finite simple group with a subgroup of index n. Show that G is then isomorphic to a subgroup of A_n. Suggestion: use Exercise 10.23.

12.14 Prove that there is no simple group of order 80.

Chapter 13

Abelian Groups

As we have seen, finite groups, even small ones, are complicated and very difficult to classify. However, abelian groups are quite a different story. As we shall see, it is not hard to classify finite abelian groups, or even finitely generated abelian groups. This is the goal of this chapter.

Free Abelian Groups

Recall that a group G is *finitely generated* if there is a finite subset $g \subset G$ such that $G = \langle g \rangle$. In this chapter, all groups will be abelian and finitely generated. As is usual in abelian groups, we shall write the group operation as addition.

DEFINITION 13.1   *A set of generators $g = \{\alpha_1, \ldots, \alpha_n\}$ of a finitely generated abelian group G is called a basis of G if there are no nontrivial relations among the elements of g. In other words,*

$$a_1 \alpha_1 + \cdots + a_n \alpha_n = 0 \,,$$

for $a_1, \ldots, a_n \in \mathbb{Z}$, implies that

$$a_1 = \cdots = a_n = 0 \,.$$

If there exists a basis for G, then G is called a free abelian group.

For example, $\mathbb{Z}^n$ has the basis $\{\epsilon_1, \ldots, \epsilon_n\}$, where

$$\epsilon_i = \left(0, \ldots, 0, \underset{i}{1}, 0, \ldots, 0 \right)$$

for $1 \leq i \leq n$, and is thus a free abelian group. On the other hand, $\mathbb{Z}/n\mathbb{Z}$ is not free, because every element α satisfies

$$n\alpha = 0 \,.$$

In fact, since a free group is infinite, no finite abelian group is free. In general, if G is an abelian group, set

$$G_t = \{\alpha \in G \mid n\alpha = 0, \text{ for some } n \in \mathbb{Z}\} \, .$$

It is easy to see that G_t is a subgroup of G, called the *torsion* subgroup of G. One consequence of the classification theorem presented later in this chapter will be that

$$G \cong G_{fr} \times G_t \, ,$$

where G_{fr} is a free subgroup of G.

Every finitely generated free abelian group is isomorphic to $\mathbb{Z}^n$ for some n. Why is this so? Suppose G has a basis $\{\alpha_1, \dots, \alpha_n\}$. Define a mapping

$$f : \mathbb{Z}^n \to G$$

by

$$f(a_1, \dots, a_n) := a_1\alpha_1 + \cdots + a_n\alpha_n \, .$$

This mapping is clearly a homomorphism and is surjective since $\alpha_1, \dots, \alpha_n$ generate G. Suppose $f(a_1, \dots, a_n) = 0$ for some $(a_1, \dots, a_n) \in \mathbb{Z}^n$. So

$$a_1\alpha_1 + \cdots + a_n\alpha_n = 0 \, .$$

Then since there are no nontrivial relations among $\alpha_1, \dots, \alpha_n$, we have $a_1 = \cdots = a_n = 0$. Thus f is injective.

It is not hard to see that for $m \neq n$, $\mathbb{Z}^m \not\cong \mathbb{Z}^n$. Therefore, any two bases of a free abelian group have the same number of elements.

DEFINITION 13.2   *The rank of a free abelian group is the number of elements in a basis of the group.*

Now let us return to an arbitrary finitely generated abelian group G with generators $g = \{\alpha_1, \dots, \alpha_n\}$. Then as above, we have a homomorphism

$$f : \mathbb{Z}^n \to G$$

given by

$$f : (a_1, \dots, a_n) \mapsto a_1\alpha_1 + \cdots + a_n\alpha_n \, , \tag{13.1}$$

which is surjective. Its kernel is a subgroup of $\mathbb{Z}^n$. By describing the kernel precisely, we shall get a description of G. The first step is the following theorem.

THEOREM 13.1
A subgroup of a free abelian group of rank n is free of rank at most n.

PROOF   We prove the theorem by induction on n. For $n = 0$ there is nothing to prove. So assume that the result holds for any subgroup of a free group of rank less than n. Let G be a free group with basis $\{\alpha_1, \ldots, \alpha_n\}$, and H a subgroup of G. Set

$$G_1 = \langle \alpha_2, \ldots, \alpha_n \rangle \ .$$

We can assume that $H \subseteq G_1$ (Why?). The inclusion $H \hookrightarrow G$ induces a homomorphism

$$g : H \to G/G_1 \cong \mathbb{Z}\alpha_1 \cong \mathbb{Z} \ .$$

Now

$$\ker g = H \cap G_1 \ ,$$

so the induced homomorphism

$$\bar{g} : H/(H \cap G_1) \to G/G_1 \cong \mathbb{Z}$$

is injective. Therefore, by Exercise 6.10, $H/H \cap G_1$ is cyclic, and in fact,

$$H/(H \cap G_1) \cong a_1 (\mathbb{Z}\alpha_1) \ ,$$

for some $a_1 \in \mathbb{Z}$, $a_1 > 0$. Pick an element $\beta_1 \in H$ such that $\bar{\beta}_1$ generates $H/H \cap G_1$. We can assume that it is of the form

$$\beta_1 = a_1\alpha_1 + \beta \in H \ ,$$

for some $\beta \in H \cap G_1$. Since the rank of G_1 is $n - 1$, by the induction assumption, $H \cap G_1$ is free and of rank at most $n - 1$. Let $\{\beta_2, \ldots, \beta_m\}$, $m \leq n$, be a basis of $H \cap G_1$. We want to show that $\{\beta_1, \beta_2, \ldots, \beta_m\}$ is a basis of H. First we check that it generates H. Given an element $\gamma \in H$, we know that

$$\bar{\gamma} = b_1\bar{\beta}_1 \in H/(H \cap G_1) \ ,$$

for some $b_1 \in \mathbb{Z}$. Therefore $\gamma - b\beta_1 \in H \cap G_1$ and so

$$\gamma - b_1\beta_1 = b_2\beta_2 + \cdots + b_m\beta_m$$

for some $b_2, \ldots b_m \in \mathbb{Z}$. Thus

$$\gamma = b_1\beta_1 + b_2\beta_2 + \cdots + b_m\beta_m \ ,$$

and $\{\beta_1, \ldots, \beta_m\}$ generates H.

Now suppose

$$b_1\beta_1 + b_2\beta_2 + \cdots + b_m\beta_m = 0 \ ,$$

for some $b_1, \ldots, b_m \in \mathbb{Z}$. Then in $H/(H \cap G_1)$,

$$b_1\bar{\beta}_1 = 0 \ ,$$

which means that $b_1 = 0$ since $H/(H \cap G_1)$ is free. But then

$$b_2\beta_2 + \cdots + b_m\beta_m = 0 .$$

Since $\{\beta_2, \ldots, \beta_m\}$ is a basis of $H \cap G_1$, we have that

$$b_2 = \cdots = b_m = 0$$

as well. Therefore $\{\beta_1, \ldots, \beta_m\}$ is a basis of H. By the principle of induction, the result then holds for all n. ∎

In particular, this shows that the kernel of the homomorphism f, defined in (13.1) above, is a free subgroup of $\mathbb{Z}^n$. Suppose $\{\beta_1, \ldots, \beta_m\} \subset \mathbb{Z}^n$ is a basis of $\ker f$. We can write

$$\beta_j = \sum_{i=1}^{n} a_{ij}\epsilon_i ,$$

for some $a_{ij} \in \mathbb{Z}$. If we let $A = (a_{ij}) \in M(n, m, \mathbb{Z})$, then we have the sequence of homomorphisms

$$\mathbb{Z}^m \xrightarrow{A} \mathbb{Z}^n \xrightarrow{f} G ,$$

with

$$\ker f = \langle \beta_1, \ldots, \beta_m \rangle = \operatorname{im} A .$$

Thus

$$G \cong \mathbb{Z}^n / \operatorname{im} A .$$

In the next section, we shall show that there is a basis $\{\alpha_1, \ldots, \alpha_n\}$ of $\mathbb{Z}^n$ such that $\{d_1\alpha_1, \ldots, d_m\alpha_m\}$ is a basis of $\ker f$, where $d_1, \ldots, d_m \in \mathbb{Z}$ and $d_1 \mid \cdots \mid d_m$. This will give us our first classification theorem.

Row and Column Reduction of Integer Matrices

Suppose that G is a free abelian group of rank n with basis $\{\alpha_1, \ldots, \alpha_n\}$, and H is a subgroup of rank $m \leq n$ with basis $\{\beta_1, \ldots, \beta_m\}$. Write

$$\beta_j = \sum_{i=1}^{n} a_{ij}\alpha_i , \tag{13.2}$$

for some $a_{ij} \in \mathbb{Z}$, $1 \leq i \leq n$, $1 \leq j \leq m$. Let $A = (a_{ij}) \in M(n, m, \mathbb{Z})$. Our goal is to diagonalize A using integral row and column operations. The algorithm in fact applies to an arbitrary $n \times m$ integer matrix. First, let us list the elementary operations.

(i) Multiply row (column) i by -1.

(ii) Interchange rows (columns) i and j.

(iii) Add a times row (column) i to row (column) j, where $a \in \mathbb{Z}$.

As with real row and column operations, these integer operations correspond to multiplication on the left or right by elementary matrices. To obtain the corresponding elementary matrix, apply the operation to the $n \times n$ or $m \times m$ identity matrix. Row operations change the basis of G, and column operations, the basis of H.

Now we diagonalize A. First, pick an entry of A of minimal size. By interchanging a row and a column, move it to the position $(1, 1)$. If necessary, multiply by -1 to make it nonnegative. Divide each entry in row 1 by a_{11}. If a remainder is not 0, then move that entry to the position $(1, 1)$. Continue until all entries in row 1, except the first, are 0. Do the same with column 1. We now have a matrix of the form

$$\begin{pmatrix} a_{11} & 0 & \cdots & 0 \\ 0 & a_{22} & \cdots & a_{2m} \\ \vdots & \vdots & \ddots & \vdots \\ 0 & a_{n2} & \cdots & a_{nm} \end{pmatrix}.$$

We also want a_{11} to divide all other entries. Suppose that there is an entry in row i that is not divisible by a_{11}. Add row i to row 1. Then proceed as before to make all other entries in row 1, 0. Continuing in this way, we end up with an entry in position $(1, 1)$ that divides all other entries in A.

Now apply the same procedure to the $(n-1) \times (m-1)$ matrix remaining. Applying row and column operations will leave the entries divisible by a_{11}. We end up with a matrix of the form

$$B = \begin{pmatrix} d_1 & 0 & \cdots & 0 \\ 0 & d_2 & \cdots & 0 \\ \vdots & \vdots & \ddots & \vdots \\ 0 & 0 & \cdots & d_m \\ 0 & 0 & \cdots & 0 \\ \vdots & \vdots & \ddots & \vdots \\ 0 & 0 & \cdots & 0 \end{pmatrix}$$

where $d_i \geq 0$, for $1 \leq i \leq m$, and

$$d_1 \mid d_2 \mid \cdots \mid d_m .$$

Furthermore,

$$B = PAQ ,$$

where $P \in M(n, \mathbb{Z})$ and $Q \in M(m, \mathbb{Z})$ are invertible. Thus we have proved:

THEOREM 13.2
Let $A \in M(n, m, \mathbb{Z})$. Then there exist invertible matrices $P \in M(n, \mathbb{Z})$ and $Q \in M(m, \mathbb{Z})$ such that $B = PA$ and Q is a diagonal matrix with nonnegative diagonal entries $d_1 \mid d_2 \mid \cdots \mid d_m$.

Now these numbers $d_1, \ldots, d_m$ are in fact invariants of A, that is they do not depend on how A is diagonalized. To see this, we will show that they can be expressed in terms of the minors of A. For any matrix $A \in M(n, m, \mathbb{Z})$, let $\delta_k = \delta_k(A)$ be the greatest common divisor of the $k \times k$ minors of A, for $1 \le k \le m$.

LEMMA 13.1
Suppose that $P \in M(n, \mathbb{Z})$ and $Q \in M(m, \mathbb{Z})$ are invertible. Then $\delta_1, \ldots, \delta_m$ are the same for A and for PAQ.

PROOF   The rows of PA are integral linear combinations of the rows of A. So for any k, $1 \le k \le m$, the $k \times k$ minors of PA are linear combinations of the $k \times k$ minors of A. Therefore, the greatest common divisor of the $k \times k$ minors of A divides all the $k \times k$ minors of PA, and thus divides their greatest common divisor. Now $A = P^{-1}(PA)$. So, reversing the roles of A and PA, the same argument shows that $\delta_k(PA)$ divides $\delta_k(A)$. Thus

$$\delta_k(PA) = \delta_k(A) .$$

Multiplying PA on the right by Q has a similar effect: the columns of PAQ are linear combinations of the columns of PA. So by an argument similar to the one we have just made, we see that

$$\delta_k(PAQ) = \delta_k(PA) .$$

Therefore,

$$\delta_k(PAQ) = \delta_k(A) ,$$

as claimed.   ∎

If $B \in M(n, m, \mathbb{Z})$ is a diagonal matrix with nonnegative diagonal entries $d_1 \mid d_2 \mid \cdots \mid d_m$, then

$$\delta_k = d_1 \cdots d_k ,$$

for $1 \le k \le m$. Equivalently,

$$d_k = \delta_k/\delta_{k-1} ,$$

for $k > 1$, provided $\delta_{k-1} \ne 0$, and $d_1 = \delta_1$. We can now prove that $d_1, \ldots, d_m$ are invariants of A.

THEOREM 13.3
Let $A \in M(n, m, \mathbb{Z})$. Suppose we diagonalize A and obtain a diagonal matrix with diagonal entries $d_1 \mid d_2 \mid \cdots \mid d_m$. If we diagonalize A in a different way, we will obtain the same diagonal matrix.

PROOF   The lemma shows that the invariants $\delta_1, \ldots, \delta_m$ are the same for A and PAQ, for any invertible $P \in M(n, \mathbb{Z})$ and $Q \in M(m, \mathbb{Z})$. Diagonalizing A means finding such P and Q so that PAQ is diagonal. So regardless of how we diagonalize A, the resulting diagonal matrices will have the same invariants $\delta_1, \ldots, \delta_m$. But as we have seen, these determine the diagonal entries of the resulting diagonal matrices.

∎

Let us now return to the matrix A given by Equation (13.2). In this case, it has rank m. The matrix P transforms our original basis $\{\alpha_1, \ldots, \alpha_n\}$ of G into a basis $\{\gamma_1, \ldots, \gamma_n\}$ of G such that

$$\{d_1\gamma_1, \ldots, d_m\gamma_m\}$$

is a basis of H. This proves the following result.

THEOREM 13.4

Let G be a free abelian group of rank n and H a subgroup. Then there exists a basis $\{\alpha_1, \ldots, \alpha_n\}$ of G and positive integers $d_1 \mid \cdots \mid d_m$, for some $m \leq n$, such that

$$\{d_1\alpha_1, \ldots, d_m\alpha_m\}$$

is a basis of H.

Example 13.1

Let

$$A = \begin{pmatrix} 0 & 2 & 0 \\ -6 & -4 & -6 \\ 6 & 6 & 6 \\ 7 & 10 & 6 \end{pmatrix}.$$

Apply the algorithm above to A:

$$\begin{pmatrix} 0 & 2 & 0 \\ -6 & -4 & -6 \\ 6 & 6 & 6 \\ 7 & 10 & 6 \end{pmatrix} \xrightarrow[\leftrightarrow col\,2]{col\,1} \begin{pmatrix} 2 & 0 & 0 \\ -4 & -6 & -6 \\ 6 & 6 & 6 \\ 10 & 7 & 6 \end{pmatrix} \xrightarrow{clear\ col\,1} \begin{pmatrix} 2 & 0 & 0 \\ 0 & -6 & -6 \\ 0 & 6 & 6 \\ 0 & 7 & 6 \end{pmatrix}.$$

Now 2 does not divide 7. So we add row 4 to row 1:

$$
\begin{pmatrix} 2 & 0 & 0 \\ 0 & -6 & -6 \\ 0 & 6 & 6 \\ 0 & 7 & 6 \end{pmatrix}
\xrightarrow[+\,row\,4]{row\,1}
\begin{pmatrix} 2 & 7 & 6 \\ 0 & -6 & -6 \\ 0 & 6 & 6 \\ 0 & 7 & 6 \end{pmatrix}
\xrightarrow[-\,3\,col\,1]{col\,2}
\begin{pmatrix} 2 & 1 & 6 \\ 0 & -6 & -6 \\ 0 & 6 & 6 \\ 0 & 7 & 6 \end{pmatrix}
\xrightarrow[\leftrightarrow\,col\,2]{col\,1}
$$

$$
\begin{pmatrix} 1 & 2 & 6 \\ -6 & 0 & -6 \\ 6 & 0 & 6 \\ 7 & 0 & 6 \end{pmatrix}
\xrightarrow[col\,1]{clear\,row\,1,}
\begin{pmatrix} 1 & 0 & 0 \\ 0 & 12 & 30 \\ 0 & -12 & -30 \\ 0 & -14 & -36 \end{pmatrix}
\xrightarrow[-\,2\,col\,2]{col\,3}
\begin{pmatrix} 1 & 0 & 0 \\ 0 & 12 & 6 \\ 0 & -12 & -6 \\ 0 & -14 & -8 \end{pmatrix}
\xrightarrow[\leftrightarrow\,col\,3]{col\,2}
$$

$$
\begin{pmatrix} 1 & 0 & 0 \\ 0 & 6 & 12 \\ 0 & -6 & -12 \\ 0 & -8 & -14 \end{pmatrix}
\xrightarrow[col\,2]{clear\,row\,2,}
\begin{pmatrix} 1 & 0 & 0 \\ 0 & 6 & 0 \\ 0 & 0 & 0 \\ 0 & -2 & 2 \end{pmatrix}
\xrightarrow[\leftrightarrow\,row\,2]{row\,4}
\begin{pmatrix} 1 & 0 & 0 \\ 0 & -2 & 2 \\ 0 & 0 & 0 \\ 0 & 6 & 0 \end{pmatrix}
\xrightarrow{-row\,2}
$$

$$
\begin{pmatrix} 1 & 0 & 0 \\ 0 & 2 & -2 \\ 0 & 0 & 0 \\ 0 & 6 & 0 \end{pmatrix}
\xrightarrow[col\,2]{clear\,row\,2,}
\begin{pmatrix} 1 & 0 & 0 \\ 0 & 2 & 0 \\ 0 & 0 & 0 \\ 0 & 0 & 6 \end{pmatrix}
\xrightarrow[\leftrightarrow\,row\,4]{row\,3}
\begin{pmatrix} 1 & 0 & 0 \\ 0 & 2 & 0 \\ 0 & 0 & 6 \\ 0 & 0 & 0 \end{pmatrix} .
$$

□

Classification Theorems

We can now state the first classification theorem for finitely generated abelian groups.

THEOREM 13.5

Let G be a finitely generated abelian group. There exist $d_1, \ldots, d_m \in \mathbb{N}$, with $d_1 \mid d_2 \mid \cdots \mid d_m$, and $r \geq 0$, such that

$$
G \cong \mathbb{Z}/d_1\mathbb{Z} \times \cdots \times \mathbb{Z}/d_m\mathbb{Z} \times \mathbb{Z}^r .
$$

PROOF   To show the existence of such a decomposition, we need only put together what we have discussed in the previous two sections. Let $\{\alpha_1, \ldots, \alpha_n\}$ be a set of generators of G. Define $f : \mathbb{Z}^n \to G$ by

$$
f : (a_1, \ldots, a_n) \mapsto a_1\alpha_1 + \cdots + a_n\alpha_n .
$$

This map is surjective and therefore

$$
G \cong \mathbb{Z}^n / \ker f .
$$

By Theorem 13.4, there exists a basis $\{\beta_1, \ldots, \beta_n\}$ of $\mathbb{Z}^n$ and natural numbers $d_1, \ldots, d_m$ with $d_1 \mid \cdots \mid d_m$, such that $\{d_1\beta_1, \ldots, d_m\beta_m\}$ is a basis of ker f. Therefore

$$G \cong \mathbb{Z}/d_1\mathbb{Z} \times \cdots \times \mathbb{Z}/d_m\mathbb{Z} \times \mathbb{Z}^r$$

with $r = n - m$. ∎

For example, let A be the matrix in Example 13.1, and let $G = \mathbb{Z}^4 / \operatorname{im} A$. Then

$$G \cong \mathbb{Z}/2\mathbb{Z} \times \mathbb{Z}/6\mathbb{Z} \times \mathbb{Z}$$

COROLLARY 13.1
The torsion subgroup

$$G_t \cong \mathbb{Z}/d_1\mathbb{Z} \times \cdots \times \mathbb{Z}/d_m\mathbb{Z} ,$$

and G/G_t is free of rank r.

This shows that r is an invariant of G. We define the *rank* of G to be the rank of G/G_t. In the next section, we shall show that $d_1, \ldots, d_m$ are also invariants, called the *elementary divisors* of G.

Suppose we want to use the elementary divisors to classify finite abelian groups of a given order d. How do we do this? The key is the two conditions:

(i) $d_1 \mid d_2 \mid \cdots \mid d_m$,

(ii) $d_1 d_2 \cdots d_m = d$.

For example, let us classify abelian groups of order 72. We have

$$72 = 2^3 \cdot 3^2 .$$

Since d_1 divides all the other elementary divisors, each prime factor of d_1 does as well. No prime factor of 72 occurs with multiplicity greater than 3. So there can be at most 3 elementary divisors. Begin with $m = 1$. The only possibility is

$$d_1 = 72 .$$

Next, consider $m = 2$. We must write

$$72 = d_1 d_2 , \quad \text{with} \quad d_1 \mid d_2 .$$

The possibilities are

$$72 = 2 \cdot 36$$
$$72 = 3 \cdot 24$$
$$72 = 6 \cdot 12 .$$

Lastly, let $m = 3$. We are looking for factorizations

$$72 = d_1 d_2 d_3 , \quad \text{where} \quad d_1 \mid d_2 \mid d_3 .$$

The only possibilities are

$$72 = 2 \cdot 6 \cdot 6$$
$$72 = 2 \cdot 2 \cdot 18 .$$

So there are six abelian groups of order 72.

One can also decompose a finite abelian group into a product of cyclic groups of prime power order. This is our second classification theorem.

THEOREM 13.6
Let G be a finitely generated abelian group. Then

$$G \cong (\mathbb{Z}/p_1^{k_1}\mathbb{Z})^{l_1} \times \cdots \times (\mathbb{Z}/p_m^{k_m}\mathbb{Z})^{l_m} \times \mathbb{Z}^r ,$$

where $p_1, \ldots, p_m$ are primes and $k_1, \ldots, k_m, l_1, \ldots, l_m \in \mathbb{N}$.

PROOF   Because of Theorem 13.5, we need only show that for any $d \in \mathbb{N}, d > 1$,

$$\mathbb{Z}/d\mathbb{Z} \cong (\mathbb{Z}/p_1^{k_1}\mathbb{Z}) \times \cdots \times (\mathbb{Z}/p_l^{k_l}\mathbb{Z}) , \tag{13.3}$$

where $p_1, \ldots, p_l$ are prime numbers. Write

$$d = p_1^{k_1} \cdots p_l^{k_l}$$

where $p_1, \ldots, p_l$ are distinct primes, and $k_1, \ldots, k_l \in \mathbb{N}$. Then by Example 5.4(ii), (13.3) holds.   ∎

Example 13.2
For comparison, let us use this theorem to list the abelian groups of order 72. Again we have

$$72 = 2^3 \cdot 3^2 .$$

The abelian groups of order 8 are

$$(\mathbb{Z}/2\mathbb{Z})^3 , \quad \mathbb{Z}/2\mathbb{Z} \times \mathbb{Z}/4\mathbb{Z} , \quad \mathbb{Z}/8\mathbb{Z} .$$

Those of order 9 are

$$(\mathbb{Z}/3\mathbb{Z})^2 , \quad \mathbb{Z}/9\mathbb{Z} .$$

So we have the six groups:

(i) $(\mathbb{Z}/2\mathbb{Z})^3 \times (\mathbb{Z}/3\mathbb{Z})^2 \cong \mathbb{Z}/2\mathbb{Z} \times (\mathbb{Z}/6\mathbb{Z})^2$

(ii) $(\mathbb{Z}/2\mathbb{Z})^3 \times \mathbb{Z}/9\mathbb{Z} \cong (\mathbb{Z}/2\mathbb{Z})^2 \times \mathbb{Z}/18\mathbb{Z}$

(iii) $\mathbb{Z}/2\mathbb{Z} \times \mathbb{Z}/4\mathbb{Z} \times (\mathbb{Z}/3\mathbb{Z})^2 \cong \mathbb{Z}/6\mathbb{Z} \times \mathbb{Z}/12\mathbb{Z}$

(iv) $\mathbb{Z}/2\mathbb{Z} \times \mathbb{Z}/4\mathbb{Z} \times \mathbb{Z}/9\mathbb{Z} \cong \mathbb{Z}/4\mathbb{Z} \times \mathbb{Z}/18\mathbb{Z} \cong \mathbb{Z}/2\mathbb{Z} \times \mathbb{Z}/36\mathbb{Z}$

(v) $\mathbb{Z}/8\mathbb{Z} \times (\mathbb{Z}/3\mathbb{Z})^2 \cong \mathbb{Z}/24\mathbb{Z} \times \mathbb{Z}/3\mathbb{Z}$

(vi) $\mathbb{Z}/8\mathbb{Z} \times \mathbb{Z}/9\mathbb{Z} \cong \mathbb{Z}/72\mathbb{Z}.$ ⬜

Invariance of Elementary Divisors

In this section, we shall show that the elementary divisors of an abelian group are invariants of the group. First, we set up the basic tool we shall use.

Let G be an abelian group. For any $a \in \mathbb{Z}$,

$$aG := \{a\alpha \mid \alpha \in G\}$$

is a subgroup of G. We are particularly interested in $p^j G$, where p is prime. Now the quotient group G/pG is naturally a vector space over the field $\mathbb{F}_p$. We just need to define scalar multiplication. Let

$$(a + p\mathbb{Z})(\alpha + pG) := a\alpha + pG .$$

This is clearly well-defined and makes G/pG into an $\mathbb{F}_p$-vector space. For example, if

$$G = (\mathbb{Z}/3\mathbb{Z}) \times (\mathbb{Z}/9\mathbb{Z}) ,$$

then

$$G/3G \cong (\mathbb{F}_3)^2 .$$

Since $p(p^j G) = p^{j+1}G$, the quotient group

$$p^j G/p^{j+1}G$$

is an $\mathbb{F}_p$-vector space as well. The key to our proof that the elementary divisors are invariants is the following lemma.

LEMMA 13.2
Let $G = \mathbb{Z}/d\mathbb{Z}$, for $d \in \mathbb{N}$. Then for p prime,

$$dim_{\mathbb{F}_p} p^j G/p^{j+1}G = \begin{cases} 0, & if\ p^{j+1} \nmid d , \\ 1, & if\ p^{j+1} \mid d . \end{cases}$$

PROOF   First suppose that $p^{j+1} \nmid d$. Then

$$\left(p^{j+1}, d\right) = p^k = \left(p^j, d\right),$$

for some $k \leq j$. Therefore in G,

$$|\bar{p}^j| = d/(p^j, d) = d/p^k = d/\left(p^{j+1}, d\right) = |\bar{p}^{j+1}|$$

(Exercise 5.6). Now $\bar{p}^j = p^j \cdot \bar{1}$ is a generator of $p^j G$. So this tells us that $\bar{p}^{j+1}$ is as well, and therefore

$$p^j G / p^{j+1} G = 0 .$$

However, if $p^{j+1} \mid d$, then $(p^{j+1}, d) = p^{j+1}$, whereas $(p^j, d) = p^j$. Thus

$$|\bar{p}^j| = d/p^j \quad \text{and} \quad |\bar{p}^{j+1}| = d/p^{j+1} .$$

Therefore

$$p^j G / p^{j+1} G \cong \mathbb{Z}/p\mathbb{Z} ,$$

and

$$dim_{\mathbb{F}_p} p^j G / p^{j+1} G = 1 . \quad \blacksquare$$

For example, if $d = 12$, then

$$G = \mathbb{Z}/12\mathbb{Z} \cong \mathbb{Z}/4\mathbb{Z} \times \mathbb{Z}/3\mathbb{Z} ,$$

and we have

$$G/2G \cong (\mathbb{Z}/12\mathbb{Z})/(2\mathbb{Z}/12\mathbb{Z}) \cong \mathbb{Z}/2\mathbb{Z} \quad \Leftrightarrow \quad dim_{\mathbb{F}_2} G/2G = 1$$

$$2G/4G \cong (2\mathbb{Z}/12\mathbb{Z})/(4\mathbb{Z}/12\mathbb{Z}) \cong \mathbb{Z}/2\mathbb{Z} \quad \Leftrightarrow \quad dim_{\mathbb{F}_2} 2G/4G = 1$$

$$4G/8G \cong (\mathbb{Z}/3\mathbb{Z})/(\mathbb{Z}/3\mathbb{Z}) \cong 0 \quad \Leftrightarrow \quad dim_{\mathbb{F}_2} 4G/8G = 0$$

$$G/3G \cong (\mathbb{Z}/12\mathbb{Z})/(3\mathbb{Z}/12\mathbb{Z}) \cong \mathbb{Z}/3\mathbb{Z} \quad \Leftrightarrow \quad dim_{\mathbb{F}_3} G/3G = 1$$

$$3G/9G \cong (\mathbb{Z}/4\mathbb{Z})/(\mathbb{Z}/4\mathbb{Z}) \cong 0 \quad \Leftrightarrow \quad dim_{\mathbb{F}_3} 3G/9G = 0 .$$

We are now ready to prove that the elementary divisors are invariants.

THEOREM 13.7
Suppose that

$$\mathbb{Z}/d_1\mathbb{Z} \times \cdots \times \mathbb{Z}/d_m\mathbb{Z} \times \mathbb{Z}^r \cong G \cong \mathbb{Z}/e_1\mathbb{Z} \times \cdots \times \mathbb{Z}/e_n\mathbb{Z} \times \mathbb{Z}^s$$

where $d_1 \mid \cdots \mid d_m$ *and* $e_1 \mid \cdots \mid e_n$. *Then* $r = s$, $m = n$, *and* $d_1 = e_1, \ldots, d_m = e_m$.

PROOF   From Corollary 13.1, we see that $r = s$ and

$$\mathbb{Z}/d_1\mathbb{Z} \times \cdots \times \mathbb{Z}/d_m\mathbb{Z} \cong G_t \cong \mathbb{Z}/e_1\mathbb{Z} \times \cdots \times \mathbb{Z}/e_n\mathbb{Z} .$$

Pick a prime number p. By Lemma 13.2 and Corollary 13.1,

$$\dim_{\mathbb{F}_p} G_t/pG_t \leq m .$$

Furthermore,

$$\dim_{\mathbb{F}_p} G_t/pG_t = m$$

if $p \mid d_1$, since this implies that $p \mid d_k$ for all k. In particular,

$$m = \max_p \dim_{\mathbb{F}_p} G_t/pG_t .$$

The same holds for n. Therefore $m = n$.

Now for any prime, and any $j \geq 0$,

$$l := \dim p^j G_t / p^{j+1} G_t$$

is the number of d_k such that p^{j+1} divides d_k. Keeping in mind that if p^{j+1} divides d_k, then it also divides $d_{k+1}, \ldots, d_m$. This tells us that

$$p^{j+1} \nmid d_1, \ldots, d_{m-l} \quad \text{but} \quad p^{j+1} \mid d_{m-l+1}, \ldots, d_m .$$

So these dimensions determine the prime factorization of $d_1, \ldots, d_m$. The same holds for $e_1, \ldots, e_m$. Therefore,

$$d_1 = e_1, \ldots, d_m = e_m . \quad \blacksquare$$

For example, suppose that G is a finite abelian group with

$$\dim_{\mathbb{F}_2} G/2G = 7 \tag{13.4}$$
$$\dim_{\mathbb{F}_2} 2G/4G = 4 \tag{13.5}$$
$$\dim_{\mathbb{F}_2} 4G/8G = 2 \tag{13.6}$$
$$\dim_{\mathbb{F}_2} 8G/16G = 1 \tag{13.7}$$
$$\dim_{\mathbb{F}_p} G/pG = 0 ,$$

for $p \neq 2$. Then we know that the number of elementary divisors is $m = 7$. Furthermore, (13.4) implies that 2 divides all of them; (13.5) that 4 divides d_4, d_5, d_6, and d_7; (13.6) that 8 divides d_6 and d_7, and (13.7) that 16 divides d_7. Therefore,

$$d_1 = d_2 = d_3 = 2 , \quad d_4 = d_5 = 4 , \quad d_6 = 8 , \quad d_7 = 16$$

So

$$G \cong (\mathbb{Z}/2\mathbb{Z})^3 \times (\mathbb{Z}/4\mathbb{Z})^2 \times \mathbb{Z}/8\mathbb{Z} \times \mathbb{Z}/16\mathbb{Z} .$$

The Multiplicative Group of the Integers Mod n

An interesting class of abelian groups is the multiplicative groups $(\mathbb{Z}/n\mathbb{Z})^\times$. How do they decompose? First, recall that if $n = p_1^{k_1} \cdots p_m^{k_m}$, with $p_1, \ldots, p_m$ distinct primes, then

$$\mathbb{Z}/n\mathbb{Z} \cong \mathbb{Z}/p_1^{k_1}\mathbb{Z} \times \cdots \times \mathbb{Z}/p_m^{k_m}\mathbb{Z}$$

(see Example 5.5(ii)). It is easy to see that

$$(\mathbb{Z}/n\mathbb{Z})^\times \cong \left(\mathbb{Z}/p_1^{k_1}\mathbb{Z}\right)^\times \times \cdots \times \left(\mathbb{Z}/p_m^{k_m}\mathbb{Z}\right)^\times .$$

(see Exercise 5.24). The question is then: for p prime, $k \in \mathbb{N}$, what does $(\mathbb{Z}/p^k\mathbb{Z})^\times$ look like?

First we introduce the *p-adic expansion* of a natural number a.

THEOREM 13.8
Any $a \in \mathbb{N}$ has a unique p-adic expansion

$$a = a_0 + a_1 p + \cdots + a_k p^k ,$$

where $0 \leq a_i < p$, for all i.

PROOF   First we show that such an expansion exists. Pick the smallest k such that $p^{k+1} > a$ and argue by induction on k. If $k = 0$ take $a_0 = a$ and we are finished. Suppose that the result holds for $k-1$, i.e., any $a < p^k$ has such an expansion. Divide a by p^k:

$$a = a_k p^k + b ,$$

where $0 \leq b < p^k$. Since $p^{k+1} > a$, we have $a_k < p$. By assumption, we can write

$$b = a_0 + \cdots + a_{k-1} p^{k-1} .$$

Therefore

$$a = a_0 + \cdots + a_{k-1} p^{k-1} + a_k p^k ,$$

and by the principle of induction, the result holds for all k. This argument also shows that k and $a_k, \ldots, a_1$ are uniquely determined and gives an algorithm for computing them.   ∎

For example, take $a = 744$ and $p = 7$. The smallest k for which $7^k > 744$ is $k = 4$. So we divide 744 by 7^3:

$$744 = 2 \cdot 7^3 + 58 .$$

Then we divide 58 by 7^2, and so on:

$$744 = 2 \cdot 7^3 + 58$$
$$= 2 \cdot 7^3 + 7^2 + 9$$
$$= 2 \cdot 7^3 + 7^2 + 7 + 2 .$$

We can describe $(\mathbb{Z}/p^k\mathbb{Z})^\times$ using p-adic expansions. Any element in $(\mathbb{Z}/p^k\mathbb{Z})^\times$ can be represented by a unique integer a, $0 < a < p^k$, which is prime to p. If we write

$$a = a_0 + a_1 p + \cdots + a_{k-1} p^{k-1} , \tag{13.8}$$

then $(a, p) = 1$ if and only if $a_0 \neq 0$. Counting such integers, we see that

$$\left| (\mathbb{Z}/p^k\mathbb{Z})^\times \right| = (p - 1)p^{k-1} .$$

Now $(\mathbb{Z}/p^k\mathbb{Z})^\times$ has a distinguished subgroup. For $p > 2$, let

$$U_{p^k} = \ker \left[(\mathbb{Z}/p^k\mathbb{Z})^\times \to (\mathbb{Z}/p\mathbb{Z})^\times \right] ,$$

where the homomorphism is reduction modulo p. These are just the elements represented by integers with a p-adic expansion (13.8) where $a_0 = 1$. Thus

$$|U_{p^k}| = p^{k-1} .$$

In the case $p = 2$, this group would coincide with $(\mathbb{Z}/p^k\mathbb{Z})^\times$ itself. The right definition in this case is

$$U_{2^k} = \ker \left[(\mathbb{Z}/2^k\mathbb{Z})^\times \to (\mathbb{Z}/4\mathbb{Z})^\times \right] .$$

So these are elements that can be represented by integers with a 2-adic expansion (13.8) where $a_0 = a_1 = 1$. Therefore,

$$|U_{2^k}| = 2^{k-2} .$$

The groups U_{p^k} are the key to finding the structure of $(\mathbb{Z}/p^k\mathbb{Z})^\times$:

THEOREM 13.9

The group U_{p^k} is cyclic. Assume $k > 1$. Then for $p > 2$, the element $\overline{1 + p}$ is a generator, and for $p = 2$, the element $\overline{5}$.

PROOF   Suppose that $p > 2$. Since $|U_{p^k}| = p^{k-1}$, the order of $\overline{1 + p}$ must be a power of p. But by the binomial formula,

$$(1 + p)^{p^{k-2}} \equiv 1 + p^{k-1} \pmod{p^k}$$
$$\not\equiv 1 \pmod{p^k} .$$

Therefore,

$$|\overline{1+p}| = p^{k-1} ,$$

and U_{p^k} is cyclic.
   If $p = 2$, we have

$$(1 + 2^2)^{2^{k-3}} \equiv 1 + 2^{k-1} \quad (\text{mod } 2^k)$$
$$\not\equiv 1 \quad (\text{mod } 2^k) .$$

Therefore,

$$|\bar{5}| = 2^{k-2} ,$$

and U_{2^k} is cyclic too.   ∎

COROLLARY 13.2
For $p > 2$, $(\mathbb{Z}/p^k\mathbb{Z})^\times$ is cyclic, and $(\mathbb{Z}/2^k\mathbb{Z})^\times \cong U_{2^k} \times (\mathbb{Z}/4\mathbb{Z})^\times$.

PROOF   For $p > 2$, we have that $|(\mathbb{Z}/p^k\mathbb{Z})^\times| = (p-1)p^{k-1}$. So decomposing the group into a product of cyclic groups of prime power order, we see that one factor is U_{p^k}, and the product of the remaining factors is isomorphic to $(\mathbb{Z}/p\mathbb{Z})^\times$. Thus

$$\left(\mathbb{Z}/p^k\mathbb{Z}\right)^\times \cong U_{p^k} \times (\mathbb{Z}/p\mathbb{Z})^\times .$$

As we shall see (see Application 14.1), $(\mathbb{Z}/p\mathbb{Z})^\times$ is cyclic. Therefore, since the orders of these two groups are relatively prime, by Example 5.5(ii), their product is cyclic.
   If $p = 2$, then U_{2^k} is cyclic of order 2^{k-2} and $(\mathbb{Z}/4\mathbb{Z})^\times$ is cyclic of order 2. So by Theorem 13.6, $(\mathbb{Z}/2^k\mathbb{Z})^\times$ is either the product of the two or is cyclic. Now let

$$a = 1 + 2 + \cdots + 2^{k-1} = 2^k - 1 .$$

Then
$$a^2 = (2^k - 1)^2 \equiv 1 \quad (\text{mod } 2^k) \quad \text{and} \quad a \equiv 3 \quad (\text{mod } 4) .$$

Therefore, $\langle \bar{a} \rangle \subset (\mathbb{Z}/2^k\mathbb{Z})^\times$ maps isomorphically onto $(\mathbb{Z}/4\mathbb{Z})^\times$ under reduction mod 4, and

$$(\mathbb{Z}/2^k\mathbb{Z})^\times \cong U_{2^k} \times \langle \bar{a} \rangle \cong U_{2^k} \times (\mathbb{Z}/4\mathbb{Z})^\times .\quad ∎$$

Exercises

   13.1 Prove that if $\mathbb{Z}^m \cong \mathbb{Z}^n$, then $m = n$.

   13.2 Let G be an abelian group. Show that G_t is a subgroup of G.

13.3 Suppose that
$$G = \mathbb{Z}/d_1\mathbb{Z} \times \cdots \times \mathbb{Z}/d_m\mathbb{Z} \times \mathbb{Z}^r ,$$
where $d_1, \ldots, d_m \in \mathbb{N}$, and $d_1, \ldots, d_m > 1$. Prove that
$$G_t = \mathbb{Z}/d_1\mathbb{Z} \times \cdots \times \mathbb{Z}/d_m\mathbb{Z} .$$

13.4 Let
$$A = \begin{pmatrix} -22 & -48 & 267 \\ -4 & -4 & 31 \\ -4 & -24 & 105 \\ 4 & -6 & -6 \end{pmatrix} .$$
Find the elementary divisors of $\mathbb{Z}^4 / \operatorname{im} A$.

13.5 What are the elementary divisors of
$$\mathbb{Z}/2\mathbb{Z} \times (\mathbb{Z}/6\mathbb{Z})^2 \times \mathbb{Z}/21\mathbb{Z} \times \mathbb{Z}/50\mathbb{Z} ?$$

13.6 Classify abelian groups of order 16.

13.7 Classify abelian groups of order 360.

13.8 • Let G be a finite abelian group, with elementary divisors $d_1 \mid \cdots \mid d_m$. Show that
$$d_m = \min\{n \in \mathbb{N} \mid n\alpha = 0, \text{ for all } \alpha \in G\} .$$

13.9 Let p be a prime number and let
$$G = \left\{ \frac{a}{p^k} \in \mathbb{Q} \mid a \in \mathbb{Z}, \ k \geq 0 \right\} .$$

(a) Verify that G is a subgroup of $\mathbb{Q}$. Is G finitely generated?

(b) Show that every element in $G/\mathbb{Z}$ has finite order, and that $G/\mathbb{Z}$ consists of precisely the elements of $\mathbb{Q}/\mathbb{Z}$ whose order is a power of p. Is $G/\mathbb{Z}$ finitely generated?

13.10 Give the 3-adic, 5-adic, and 7-adic expansions of 107.

13.11 Write the elements of U_{27} as powers of $\bar{4}$.

13.12 Let G be a group. Let G' be the subgroup generated by all commutators, that is, elements of the form
$$\alpha\beta\alpha^{-1}\beta^{-1} ,$$
for $\alpha, \beta \in G$. Show that G' is a normal subgroup (called the commutator subgroup), and that G/G' is abelian. Prove that if $K \subset G$ is a normal subgroup, such that G/K is abelian, then $K \supset G'$.

13.13 Write a *Mathematica* function that diagonalizes an integer matrix.

Chapter 14

Polynomial Rings

In the coming chapters, we are going to use the group theory discussed so far to see how to solve polynomial equations. What "solving" an equation means is a rather delicate question. People have known how to write solutions for a quadratic equation in terms of the square root of its discriminant for some 4000 years, and today everyone learns the formula in high school. It is simple and very useful. In the Renaissance, similar formulas were discovered for cubics and quartics. However, they are much more complicated and much less useful. Early in the 19th century, it was realized that for equations of degree greater than 4, formulas do not even exist for solutions in terms of radicals. At the same time, several mathematicians noticed that the symmetries of an equation, as we discussed in some examples in Chapter 7, tell you many interesting and profound things about its solutions. This point of view has been developed with great success in the past two centuries and will be the theme of the remainder of this book. If you are interested in the history of these ideas, the first part of van der Waerden's *History of Algebra* [9] is a good reference.

To begin, we set out the basic properties of polynomials. Then we clarify what we mean by "algebraic relations" among the roots of a polynomial. To do this we introduce field extensions, in particular the splitting field of a polynomial. With this apparatus in place, we can explain exactly what the symmetry group or Galois group of an equation is, and what its properties are. We shall give two classical applications of this theory: first, to prove that an equation of degree greater than 4 cannot in general be solved by taking roots, and secondly, to discuss geometric constructions with straight edge and compass.

In this chapter, we will look at polynomials with coefficients in an arbitrary field. These behave in many ways like the integers. There is a Euclidean algorithm for long division. There are "prime" polynomials and there is unique factorization into "primes". And the set of all polynomials with coefficients in a given field has formal properties like those of $\mathbb{Z}$.

Basic Properties of Polynomials

To begin with, let F be a field. A polynomial with coefficients in F is an expression of the form

$$f(x) = a_m x^m + \ldots + a_1 x + a_0 ,$$

where $a_0, \ldots, a_m \in F$. We define the *degree* of the polynomial f, written $\deg f$, to be the degree of the highest monomial with a nonzero coefficient:

$$\deg f = \max\{n \mid a_n \neq 0\} .$$

If $f(x) = a_n x^n + \ldots + a_1 x + a_0$ with $a_n \neq 0$, then a_n is called the *leading coefficient* of f. We can add polynomials in the obvious way:

$$(a_m x^m + \ldots + a_1 x + a_0) + (b_m x^m + \ldots + b_1 x + b_0)$$
$$= (a_m + b_m) x^m + \ldots + (a_1 + b_1) x + (a_0 + b_0)$$

and multiply them:

$$\left(a_m x^m + \ldots + a_1 x + a_0\right) \left(b_m x^m + \ldots + b_1 x + b_0\right) = c_{2m} x^{2m} + \ldots + c_1 x + c_0$$

where

$$c_n = \sum_{i=0}^{n} a_i b_{n-i} .$$

(The two polynomials can have any degree $\leq m$. It makes it easier to write the formulas if you allow terms with zero coefficients.) Clearly, for any f, g,

$$\deg(fg) = \deg f + \deg g .$$

The set of all polynomials with coefficients in F, we will denote by $F[x]$:

$$F[x] = \left\{ a_m x^m + a_{m-1} x^{m-1} + \ldots + a_1 x + a_0 \mid m \geq 0, \ a_0, \ldots, a_m \in F \right\} .$$

We can regard F as the set of constant polynomials in $F[x]$.

Next, we show that we can use long division with polynomials. Remember that for $a, b \in \mathbb{Z}$, $a, b \neq 0$,

$$b = qa + r ,$$

where $q, r \in \mathbb{Z}$, and $0 \leq r < |a|$. Here is the analogous statement for polynomials.

THEOREM 14.1
Suppose $f, g \in F[x]$, $f, g \neq 0$. Then there exist unique $q, r \in F[x]$, with $\deg r < \deg f$, such that

$$g = qf + r .$$

PROOF   Let $m = \deg f$ and $n = \deg g$. We will argue by induction on $n - m$. If $n - m < 0$, in other words $\deg g < \deg f$, then we take $q = 0$ and $r = g$. Now let $n - m = l \geq 0$ and assume that the statement of the theorem holds for $n - m < l$. Suppose that a_m is the leading coefficient of f, and b_n the leading coefficient of g. Then we can write

$$g(x) = (b_n/a_m)\, x^{n-m} f(x) + h(x)$$

where $\deg h < \deg g$. Therefore $\deg h - \deg f < l$. So by the induction assumption there exist $q_1, r \in F[x]$, with $\deg r < \deg f$ such that

$$h(x) = q_1(x) f(x) + r(x) \,.$$

But then

$$g(x) = \big((b_n/a_m)x^{n-m} + q_1(x)\big) f(x) + r(x) \,.$$

Thus the statement holds for $n - m = l$ and by the principle of induction, for all values of $n - m$.

To see that q and r are unique, suppose that there exist q' and r' as well, such that

$$g = q'f + r' \,,$$

with $\deg r' < \deg f$. Then

$$(q - q')f = r' - r \,.$$

But $\deg(r' - r) < \deg(q - q')f$ unless $q - q' = 0$. But then $r' = r$ too. So q and r are uniquely determined.   ∎

Now we can define common divisors just as we did for the integers in Chapter 1. If $f, g \in F[x]$, then one says that f *divides* g, and writes

$$f \mid g \,,$$

if $g = qf$ for some $q \in F[x]$. Notice that a nonzero scalar $a \in F^\times$ divides any polynomial $g \in F[x]$. A polynomial d is a *common divisor* of f and g if $d \mid f$ and $d \mid g$. In order to have a unique greatest common divisor, we make the following definition: a polynomial $d \in F[x]$ is *monic* if its leading coefficient is 1.

DEFINITION 14.1   *The greatest common divisor of $f, g \in F[x]$ is the common divisor of f and g, which is monic and of greatest degree.*

As for integers, the greatest common divisor is denoted by (f, g). Just as for integers, the greatest common divisor can be computed using the *Euclidean algorithm*.

We write

$$g = qf + r \qquad\qquad \deg r < \deg f$$
$$f = q_1 r + r_1 \qquad\qquad \deg r_1 < \deg r$$
$$\vdots \qquad\qquad\qquad \vdots$$
$$r_{i-1} = q_{i+1} r_i + r_{i+1} \qquad\qquad \deg r_{i+1} < \deg r_i$$
$$\vdots \qquad\qquad\qquad \vdots$$
$$r_{n-2} = q_n r_{n-1} + r_n \qquad\qquad \deg r_n < \deg r_{n-1}$$
$$r_{n-1} = q_{n+1} r_n \; ,$$

for some n. To see that this algorithm computes (f, g) we argue just as in Chapter 1. First we prove:

LEMMA 14.1
Let u and v be polynomials in $F[x]$, not both 0. Write

$$u = qv + r \; ,$$

for some q and r with $\deg r < \deg v$. *Then*

$$(u, v) = (v, r) \; .$$

The proof is the same as the proof of Lemma 1.1. Applying this to the list of divisions above we obtain

$$(r_{i-1}, r_i) = (r_i, r_{i+1})$$

for each $i < n$. Now the last equation says that $r_n \mid r_{n-1}$. This means that

$$r_n = a \, (r_{n-1}, r_n) \; ,$$

for some $a \in F^\times$. Therefore arguing by induction,

$$r_n = a \, (r_{i-1}, r_i)$$

for some $a \in F^\times$, for any i, in particular

$$r_n = a(f, g) \; .$$

So up to a scalar factor, r_n is the greatest common divisor of f and g. It is easy to see that any common divisor of f and g divides (f, g).

As in Chapter 1, we can read more out of this list of equations. The first equation can be rewritten

$$r = g - qf \; .$$

Using this, we can rewrite the second one:

$$r_1 = f - q_1 r = f - q_1(g - qf) = (1 + q_1 q)f - q_1 g \; .$$

In other words, r and then r_1 are linear combinations of f and g, with coefficients from $F[x]$. The third equation shows that r_2 is a linear combination of r_1 and r, and therefore of f and g. Continuing like this, we find that r_n is a linear combination of f and g. Thus there exist $s, t \in F[x]$ such that

$$(f, g) = sf + tg .$$

Example 14.1
In $\mathbb{F}_{11}[x]$, let

$$f(x) = x^4 + x^3 + x^2 + 3x + 2$$

and

$$g(x) = x^5 - x^4 - x^3 + 2x^2 - x - 2 .$$

Then

$$x^5 - x^4 - x^3 + 2x^2 - x - 2 = (x - 2)(x^4 + x^3 + x^2 + 3x + 2) + (x^2 + 3x + 2)$$
$$x^4 + x^3 + x^2 + 3x + 2 = (x^2 - 2x + 5)(x^2 + 3x + 2) + 3(x + 1)$$
$$x^2 + 3x + 2 = (4x + 8)(3x + 3) .$$

Therefore

$$x + 1 = (x^4 + x^3 + x^2 + 3x + 2, \ x^5 - x^4 - x^3 + 2x^2 - x - 2) .$$

Furthermore

$$x^2 + 3x + 2 = (x^5 - x^4 - x^3 + 2x^2 - x - 2) - (x - 2)(x^4 + x^3 + x^2 + 3x + 2)$$
$$x + 1 = 4(x^4 + x^3 + x^2 + 3x + 2) - 4(x^2 - 2x + 5)(x^2 + 3x + 2)$$
$$= 4(x^4 + x^3 + x^2 + 3x + 2) - (4x^2 + 3x + 9)$$
$$\left[(x^5 - x^4 - x^3 + 2x^2 - x - 2) - (x - 2)(x^4 + x^3 + x^2 + 3x + 2) \right]$$
$$= (4x^3 + 6x^2 + 3x + 8)(x^4 + x^3 + x^2 + 3x + 2)$$
$$+ (7x^2 + 8x + 2)(x^5 - x^4 - x^3 + 2x^2 - x - 2) .$$

So we can take

$$s = 4x^3 + 6x^2 + 3x + 8 \qquad \text{and} \qquad t = 7x^2 + 8x + 2 . \qquad \Box$$

We say that f and g are *relatively prime* if $(f, g) = 1$, that is, if they have no common divisors except the nonzero scalars. Thus, if f and g are relatively prime, there exist polynomials s and t such that

$$1 = sf + tg .$$

For example, $x^2 + 1$ and $x + 1$ in $\mathbb{Q}[x]$ are relatively prime and

$$1 = \frac{1}{2}\left(x^2 + 1\right) - \frac{1}{2}(x - 1)(x + 1) .$$

We also want to discuss roots of polynomials and their relation to divisors.

DEFINITION 14.2   *A root of a polynomial $f \in F[x]$ is an element $a \in F$ such that $f(a) = 0$.*

THEOREM 14.2
$a \in F$ is a root of $f \in F[x]$ if and only if $x - a$ divides f. If $\deg f = n$, then f has at most n roots in F.

PROOF   Given $a \in F$, divide f by $x - a$:

$$f = q(x - a) + r ,$$

where $\deg r < \deg(x - a)$, in other words $r \in F$. It follows that

$$f(a) = q(a - a) + r = r .$$

So a is a root of f if and only if $r = 0$, which is the case if and only if $(x - a) \mid f$.
    We can now argue by induction that if $\deg f = n$, then f has at most n roots. Start with $n = 1$. A linear polynomial $ax + b$ has one root: $-b/a$. Suppose we know that a polynomial of degree $n - 1$ has at most $n - 1$ roots. Let a be a root of f. Then

$$f = q(x - a) ,$$

where $\deg q = n - 1$. If b is any root of f, then

$$0 = f(b) = q(b)(b - a) .$$

So either b is a root of q, or $b = a$. By assumption, q has at most $n - 1$ roots. Therefore f has at most n roots. Applying the principle of induction, the result holds then for all n.   ∎

This result has a surprising application.

APPLICATION 14.1
The multiplicative group $F^\times$ of a finite field F is cyclic.

PROOF   Set $n = |F|$. So $F^\times$ is an abelian group of order $n - 1$. By Theorem 13.5

$$F^\times \cong \mathbb{Z}/d_1\mathbb{Z} \times \cdots \times \mathbb{Z}/d_m\mathbb{Z} ,$$

where $d_1, \ldots, d_m \in \mathbb{N}$, and $d_1 \mid d_2 \mid \cdots \mid d_m$. As pointed out in Exercise 13.8,

$$a^{d_m} = 1$$

for all $a \in F^\times$. Thus all $n - 1$ elements of $F^\times$ are roots of the polynomial $x^{d_m} - 1 \in F[x]$. It follows from the theorem above that

$$n - 1 \leq d_m .$$

On the other hand, since there is an element of order d_m in $F^\times$

$$d_m \leq n - 1 ,$$

by Corollary 10.2. Therefore $d_m = n - 1$ and $F^\times$ is cyclic. ∎

REMARK 14.1   The proof does not need the full power of the classification theorem for finite abelian groups. An argument using Exercise 5.8 is given in Exercise 13.16. ∎

Unique Factorization into Irreducibles

Continuing the analogy between $\mathbb{Z}$ and $F[x]$, we now explain what "primes" are in $F[x]$ and show that every polynomial can be factored uniquely into a product of "primes."

DEFINITION 14.3   *A polynomial $f \in F[x]$ is reducible if it can be factored $f = gh$, where $g, h \in F[x]$, and $\deg g, \deg h > 0$. A polynomial $f \in F[x]$ is irreducible if it is not reducible.*

In other words, a polynomial is irreducible if its only divisors are itself and the nonzero scalars. For example, $x^2 + 1 \in \mathbb{Q}[x]$ is irreducible. However, regarded as a polynomial in $\mathbb{F}_5[x]$, it is reducible because $x^2 + 1 = (x + 2)(x + 3) \in \mathbb{F}_5[x]$. A polynomial with a root is reducible. But a polynomial may be reducible without having a root. For example, in $\mathbb{Q}[x]$,

$$x^4 - 4 = (x^2 - 2)(x^2 + 2) ,$$

is reducible. But it has no roots in $\mathbb{Q}$ because neither $x^2 + 2$ nor $x^2 - 2$ have any roots in $\mathbb{Q}$.

Irreducible polynomials are analogous to prime numbers and reducible polynomials to composite numbers. Every polynomial can be written in a unique way as a product of irreducibles. The key to proving this is the following lemma.

LEMMA 14.2
Let $p \in F[x]$ be irreducible. Suppose $p \mid fg$, where $f, g \in F[x]$. Then $p \mid f$ or $p \mid g$.

PROOF Suppose that p does not divide f. Then $(p, f) = 1$ since the only monic divisor of p of degree greater than 0 is p itself. Therefore

$$1 = sp + tf ,$$

for some $s, t \in F[x]$. Multiply by g:

$$g = spg + tfg .$$

Now $p \mid spg$ and $p \mid tfg$. So $p \mid g$. ∎

It is not hard to extend this result to a product of more than two polynomials: if p is irreducible, and $p \mid f_1 \cdots f_r$, then $p \mid f_i$ for some i, $1 \leq i \leq r$. The following theorem is the analogue of the fundamental theorem of arithmetic (see [1, §2.2]) and is proved in the same way.

THEOREM 14.3
Let $f \in F[x]$, where F is a field. Then

$$f = ap_1 \cdots p_r ,$$

where $a \in F^\times$ and $p_1, \ldots, p_r \in F[x]$ are irreducible monic polynomials. This decomposition is unique up to the order of $p_1, \ldots, p_r$.

PROOF First we prove the existence of such a decomposition into irreducibles. We proceed by induction on $n := \deg f$. Linear polynomials are irreducible. So the result holds for them. Assume that it holds for all polynomials of degree less than n. If f is irreducible, then $f = ap$, where $a \in F^\times$ and p is irreducible and monic. If f is reducible, then

$$f = gh ,$$

where $\deg g, \deg h < n$. By assumption then

$$g = bp_1 \cdots p_j, \qquad h = cp_{j+1} \cdots p_r ,$$

where $b, c \in F^\times$ and $p_1, \ldots, p_r$ are irreducible monic polynomials. It follows that

$$f = (bc)p_1 \cdots p_r ,$$

as desired. So by the principle of induction, any $f \in F[x]$ can be decomposed into a product of irreducibles.

Next we demonstrate that such a decomposition is unique up to the order of the factors. Suppose that

$$f = ap_1 \cdots p_r ,$$

and

$$f = bq_1 \cdots q_s ,$$

where $a, b \in F^{\times}$ and $p_1, \ldots, p_r, q_1, \ldots, q_s \in F[x]$ are irreducible monic polynomials. For any i, $1 \leq i \leq r$, we have that

$$p_i \mid q_1 \cdots q_s .$$

Therefore by Lemma 14.2, there exists a $j(i)$, $1 \leq j(i) \leq s$, such that $p_i \mid q_{j(i)}$. But $q_{j(i)}$ is irreducible and monic. Therefore $p_i = q_{j(i)}$. Similarly, for any j, $1 \leq j \leq s$, there exists an $i(j)$, $1 \leq i(j) \leq r$, with $q_j = p_{i(j)}$. Thus $r = s$ and the factors $q_1, \ldots, q_r$ are just $p_1, \ldots, p_r$ reordered by the permutation $i \mapsto j(i)$. It follows that $a = b$ as well. This completes the proof of the theorem. ∎

Finding Irreducible Polynomials

Suppose you want to factor a polynomial in $F[x]$. You have to know which polynomials are irreducible. Deciding whether one is irreducible or not is usually not easy. In this section, we will look at two simple criteria for irreducibility of polynomials in $\mathbb{Q}[x]$, and how to list irreducible polynomials in $\mathbb{F}_p[x]$. In the last section of the chapter, we will discuss an algorithm for factoring polynomials in $\mathbb{F}_p[x]$. It will also give us a test for irreducibility.

Let us begin with

$$f(x) = a_n x^n + \cdots + a_1 x + a_0 \in \mathbb{Q}[x] .$$

If we multiply f by a common multiple a of the denominators of $a_0, \ldots, a_n$, then af has integer coefficients. One can show that af can be written as a product of integer polynomials of positive degree, if and only if f is reducible in $\mathbb{Q}[x]$.

LEMMA 14.3
Suppose $f(x) = a_n x^n + \cdots + a_1 x + a_0 \in \mathbb{Q}[x]$ with $a_0, \ldots, a_n \in \mathbb{Z}$. If f is reducible, then $f = gh$ where g and h have integer coefficients; g and h can be taken to be monic if f is monic.

PROOF   Suppose that

$$f = gh ,$$

$g, h \in \mathbb{Q}[x]$. Let b (respectively c) be a common multiple of the denominators of the coefficients of g (respectively h). Set $d = bc$. Then

$$df = g_1 h_1 ,$$

where g_1 and h_1 have integer coefficients. Now let p be a prime factor of d, and reduce this equation modulo p. We obtain

$$0 = \bar{g}_1 \bar{h}_1 \in \mathbb{F}_p[x] .$$

Therefore

$$\bar{g}_1 = 0 \qquad \text{or} \qquad \bar{h}_1 = 0 .$$

Suppose that $\bar{g}_1 = 0$. This means that all the coefficients of g_1 are divisible by p. So we can divide d and g_1 by p:

$$(d/p)f = g_2 h_2 ,$$

where g_2 and h_2 have integer coefficients. We can continue in this way with each prime factor of d until we end up with a factorization of f into a product of polynomials with integer coefficients. The leading coefficient of f is the product of the leading coefficients of g and h. So if f is monic, then the leading coefficients of g and h are both 1 or both -1. ∎

The *Eisenstein criterion* then gives a condition for a polynomial with integer coefficients to be irreducible.

THEOREM 14.4
Let

$$f(x) = x^n + \cdots + a_1 x + a_0 \in \mathbb{Q}[x] ,$$

with $a_0, \ldots, a_{n-1} \in \mathbb{Z}$. Suppose that for some prime $p \in \mathbb{Z}$,

$$p \mid a_0, \ldots, p \mid a_{n-1} ; \qquad p^2 \nmid a_0 .$$

Then f is irreducible.

PROOF   Suppose that f is reducible. By Lemma 14.3, we can assume that $f = gh$, where

$$g = x^r + \cdots b_1 x + b_0 , \qquad h = x^s + \cdots + c_1 x + c_0 ,$$

with $r + s = n$, $r, s < n$ and $b_i, c_j \in \mathbb{Z}$ for $0 \leq i \leq r - 1$, $0 \leq j \leq s - 1$. Then we have that

$$p \mid a_0 = b_0 c_0 ,$$

which implies that $p \mid b_0$ or $p \mid c_0$. It cannot divide both since by assumption $p^2 \nmid a_0$. Suppose that $p \nmid b_0$. Now reduce these polynomials modulo p: in $\mathbb{F}_p[x]$ we have

$$x^n = \bar{f}(x) = \bar{g}(x)\bar{h}(x) .$$

We have just said that $\bar{b}_0 \neq 0$ and $\bar{c}_0 = 0$. We want to show that $\bar{c}_k = 0$ for all k. Suppose we know that $\bar{c}_0 = \cdots = \bar{c}_{k-1} = 0$. Since

$$0 = \bar{a}_k = \bar{b}_k \bar{c}_0 + \cdots + \bar{b}_1 \bar{c}_{k-1} + \bar{b}_0 \bar{c}_k ,$$

it follows that

$$0 = \bar{a}_k = \bar{b}_0 \bar{c}_k ,$$

which implies that $\bar{c}_k = 0$. So by the principle of induction, $\bar{c}_k = 0$ for all k, and $\bar{h}(x) = x^s$. But then calculating the coefficient of x^s in $\bar{f}$, we see that

$$0 = \bar{a}_s = \bar{b}_0 \neq 0 ,$$

which is impossible. So f is irreducible. ∎

This criterion shows, for example, that $x^2 + 2x + 2$ is irreducible in $\mathbb{Q}[x]$. Here is a less obvious example.

Example 14.2
Let $f(x) = x^{p-1} + \cdots + x + 1 \in \mathbb{Q}[x]$, where p is a prime number. Since

$$f(x) = \frac{x^p - 1}{x - 1} ,$$

its roots in $\mathbb{C}$ are just the roots of unity other than 1 (see Example 6.3(iii)). The Eisenstein criterion does not apply directly to f. But if we make the substitution

$$x = y + 1 ,$$

then

$$g(y) := f(y+1) = \frac{(y+1)^p - 1}{y} = y^{p-1} + py^{p-2} + \cdots + \binom{p}{k} y^{k-1} + \cdots + p .$$

Since $\binom{p}{k} \equiv 0 \pmod{p}$, for $1 \leq k \leq p - 1$, (see Exercise 1.4), the criterion does apply to g. And if g is irreducible, then so is f. ⬚

A second test is based on the following observation. Let
$f(x) = a_n x^n + \cdots + a_1 x + a_0$ be a polynomial with integer coefficients. Suppose that

$$f = gh ,$$

where g and h also have integer coefficients, and $\deg g, \deg h > 0$. If we pick a prime p that does not divide the leading coefficient a_n and reduce this equation modulo p, then we obtain

$$\bar{f} = \bar{g}\bar{h} \in \mathbb{F}_p[x] .$$

Since $p \nmid a_n$, p does not divide the leading coefficients of g and h. Therefore

$$\deg \bar{g} = \deg g > 0 , \quad \deg \bar{h} = \deg h > 0 .$$

So $\bar{f}$ is reducible in $\mathbb{F}_p[x]$. Taking the converse of this gives us a test for irreducibility:

TEST 14.1

Let

$$f(x) = a_n x^n + \cdots + a_1 x + a_0 \in \mathbb{Q}[x] ,$$

where $a_0, \ldots, a_n \in \mathbb{Z}$. If there exists a prime $p \nmid a_n$, such that $\bar{f}$, the reduction of f mod p, is irreducible, then f is irreducible in $\mathbb{Q}[x]$.

This is a very practical test because it is easy to check whether polynomials in $\mathbb{F}_p[x]$ are irreducible, as we shall see in the final section of the chapter.

Example 14.3

Take $f(x) = x^5 - 5x + 12 \in \mathbb{Q}[x]$. If we reduce f modulo 7, it is not hard to check that $\bar{f} \in \mathbb{F}_7[x]$ is irreducible. Therefore f is irreducible.   ▯

You can build a list of irreducible polynomials in $\mathbb{F}_p[x]$ by using a sieve, like Eratosthene's sieve for finding prime numbers (see [1, p. 14]). First write down all the linear polynomials, then the quadratic ones, and so on. Cross out the multiples of x, of $x + 1, \ldots$, then of the remaining quadratics, ... The polynomials that are left are irreducible. It is enough to find the monic irreducibles since the others will be scalar multiples of them. For example, take $p = 2$. First list the monic polynomials over $\mathbb{F}_2$:

$$x, \; x + 1$$
$$x^2, \; x^2 + 1, \; x^2 + x, \; x^2 + x + 1$$
$$x^3, \; x^3 + 1, \; x^3 + x, \; x^3 + x^2, \; x^3 + x + 1, \; x^3 + x^2 + 1, \; x^3 + x^2 + x, \; x^3 + x^2 + x + 1$$
$$x^4, \; x^4 + 1, \; x^4 + x, \; x^4 + x^2, \; x^4 + x^3, \; x^4 + x + 1, \; \ldots$$

$$\vdots$$

Cross out multiples of the linear polynomials:

$$x, \; x + 1$$
$$x^2 + x + 1$$
$$x^3 + x^2 + 1, \; x^3 + x + 1$$
$$x^4 + x + 1, \; x^4 + x^2 + 1, \; x^4 + x^3 + 1, \; x^4 + x^3 + x^2 + x + 1$$

$$\vdots$$

Cross out the multiples of the remaining quadratics:

$$x, \ x+1$$
$$x^2 + x + 1$$
$$x^3 + x^2 + 1, \ x^3 + x + 1$$
$$x^4 + x + 1, \ x^4 + x^3 + 1, \ x^4 + x^3 + x^2 + x + 1$$
$$\vdots$$

Cross out the multiples of the remaining cubics ... , and so on. The list above already gives us the monic irreducible polynomials of degree less than 5.

You can also prove that there are infinitely many irreducible monic polynomials in $\mathbb{F}_p[x]$ by imitating the classical proof that there are infinitely many prime numbers (see [1, Theorem 1.6]). Suppose that there were only finitely many irreducible monic polynomials. Make a list of them: $f_1, f_2, \ldots, f_m$. Let

$$f = f_1 f_2 \cdots f_m + 1 \ .$$

If f were reducible, then one of the list of irreducible polynomials would divide it, say $f_j \mid f$ for some $j, 1 \leq j \leq m$. Then

$$f_j \mid (f - f_1 f_2 \cdots f_m) = 1 \ .$$

This is impossible. So f must be irreducible. It is monic since $f_1, f_2, \ldots, f_m$ are monic. But it does not occur in the list because

$$\deg f > \deg f_j$$

for all $j, 1 \leq j \leq m$. So there cannot be only finitely many monic irreducible polynomials in $\mathbb{F}_p[x]$.

Commutative Rings

We have been emphasizing similarities between $F[x]$ and $\mathbb{Z}$. The most basic similarity is that addition and multiplication look the same in both. This suggests that it is useful to make a definition that sets out these common properties.

DEFINITION 14.4   *A ring R is a set with two binary operations, "addition" and "multiplication" satisfying:*

(i) R is an abelian group under addition;

(ii) multiplication is associative;

(iii) there is an identity element for multiplication, written 1, *which is not* 0;

(iv) multiplication is distributive over addition.

A ring is called *commutative* if its multiplication is commutative. Thus $F[x]$ and $\mathbb{Z}$ are commutative rings. The set $M(n, F)$ of all $n \times n$ matrices with coefficients in a field F is a ring under matrix addition and multiplication that is not commutative. Any field is a commutative ring. In fact a field is just a commutative ring in which every nonzero element has a multiplicative inverse. In general, the *group of units* of a ring R is the set

$$R^\times = \{a \in R \mid a \text{ has a multiplicative inverse}\}$$

with the operation of ring multiplication. Thus

$$\mathbb{Z}^\times = \{\pm 1\}$$

and

$$F[x]^\times = F^\times .$$

As we saw in Chapter 1, the integers mod n, $\mathbb{Z}/n\mathbb{Z}$, have a well-defined multiplication that satisfies the properties above. So $\mathbb{Z}/n\mathbb{Z}$ is also a commutative ring. Its group of units, $(\mathbb{Z}/n\mathbb{Z})^\times$, was introduced in Example 5.1(vi) and studied in detail in Chapter 13, pages 172-174.

REMARK 14.2   Most commutative rings do not have unique factorization into primes, like $\mathbb{Z}$ and $F[x]$.   ∎

If R and S are rings, then a mapping $\psi : R \to S$ is a *ring homomorphism* if it is a group homomorphism that respects multiplication:

(i) $\psi(a + b) = \psi(a) + \psi(b)$,

(ii) $\psi(ab) = \psi(a)\psi(b)$,

(iii) $\psi(1) = 1$,

for any $a, b \in R$. For example, the canonical map $\mathbb{Z} \to \mathbb{Z}/n\mathbb{Z}$ is a ring homomorphism. A homomorphism that is bijective is called an isomorphism.

REMARK 14.3

(i) Suppose that $\psi : R \to S$ is a ring homomorphism, and let $a \in R$ be a unit. Then $\psi(a)$ is a unit in S:

$$1 = \psi(1) = \psi(aa^{-1}) = \psi(a)\psi(a^{-1}) .$$

In particular, $\psi(a) \neq 0$. Now if R is a field, then every nonzero element is a unit. So in this case, $\ker \psi = 0$ and ψ is injective.

(ii) Let F be a field. Define $\psi : \mathbb{Z} \to F$ by

$$\psi(n) := \underbrace{1 + \cdots + 1}_{n} , \qquad \psi(-n) := -\psi(n) , \qquad \psi(0) := 0 ,$$

for $n \in \mathbb{N}$. Then ψ is a ring homomorphism, and

$$\ker \psi = p\mathbb{Z} ,$$

where p is either 0 or the least positive integer in $\ker \psi$ (see Exercise 6.10).

In the first case, ψ is injective, and $\psi(n)$ has a multiplicative inverse if $n \neq 0$. Therefore we can extend ψ to all of $\mathbb{Q}$ by setting

$$\psi(m/n) = \psi(m)\psi(n)^{-1} ,$$

for any $m, n \in \mathbb{Z}$, $n \neq 0$. It is easy to see that this is a homomorphism, and by the previous remark, it must be injective. So there is a copy of $\mathbb{Q}$ inside F. Examples of such fields are $\mathbb{R}$ and $\mathbb{C}$.

The other possibility is that $p > 0$. Suppose that p is a composite number, say

$$p = qr ,$$

where $q, r > 0$. Then in F,

$$0 = p \cdot 1 = qr \cdot 1 = (q \cdot 1)(r \cdot 1) .$$

Since F is a field, we must have that either $q \cdot 1 = 0$ or $r \cdot 1 = 0$. But we chose p to be the least positive integer in $\ker \psi$. So this cannot happen, and p must be prime. Now from Theorem 10.4, we know that ψ induces a group homomorphism

$$\mathbb{F}_p = \mathbb{Z}/p\mathbb{Z} \xrightarrow{\ \bar{\psi}\ } F .$$

It is easy to see that $\bar{\psi}$ is a ring homomorphism, and therefore by the previous remark, must be injective. So in this case, F contains a copy of $\mathbb{F}_p$. An example is the field $\mathbb{F}_{p^2}$ (see Exercise 1.21), which contains $\mathbb{F}_p$ as the set of diagonal matrices. ∎

DEFINITION 14.5 *Let F be a field. If there exists a prime p such that $p \cdot 1 = 0$ in F, then p is called the characteristic of F, written*

$$\mathrm{char}\, F = p .$$

If no such p exists, then

$$\mathrm{char}\, F := 0 .$$

The copy of $\mathbb{Q}$ in F, if $\mathrm{char}\, F = 0$, or of $\mathbb{F}_p$, if $\mathrm{char}\, F = p$, is called the *prime* field of F.

Later in the book, we shall be very interested in automorphisms of a field. An *automorphism* of a field F is an isomorphism of F to itself. The set of all automorphisms forms a group under composition (see Exercise 14.11).

Just as you can construct the field of rational numbers from the integers, so can you construct the field of rational functions $F(x)$ from $F[x]$. A *rational function* over F is a quotient f/g, where $f, g \in F[x]$ and $g \neq 0$. We identify f/g with kf/kg for any $k \in F[x]$. You can define addition and multiplication just as for rational numbers:

$$\frac{f_1}{g_1} + \frac{f_2}{g_2} := \frac{f_1 g_2 + f_2 g_1}{g_1 g_2} , \qquad \frac{f_1}{g_1} \cdot \frac{f_2}{g_2} := \frac{f_1 f_2}{g_1 g_2} ,$$

where $f_1, f_2, g_1, g_2 \in F[x]$, $g_1, g_2 \neq 0$. With these two operations, $F(x)$ is a commutative ring. The ring of polynomials $F[x]$ can be regarded as a subring by identifying $f \in F[x]$ with the quotient $f/1 \in F(x)$. Any rational function $f/g \neq 0$ has a multiplicative inverse, g/f. So just like $\mathbb{Q}$, $F(x)$ is a field.

Congruences

We can also look at "congruences" modulo a polynomial and "quotient rings" analogous to $\mathbb{Z}/n\mathbb{Z}$. Suppose $f \in F[x]$ and define the subgroup (f) by

$$(f) := f F[x] = \{ fg \mid g \in F[x] \} .$$

The quotient group $F[x]/(f)$ has a well-defined multiplication induced by the multiplication on $F[x]$: given $f_1, f_2 \in F[x]$

$$(f_1 + fg_1)(f_2 + fg_2) = f_1 f_2 + f (f_1 g_2 + g_1 f_2 + fg_1 g_2) ,$$

where $g_1, g_2 \in F[x]$, so that in $F[x]/(f)$

$$\overline{(f_1 + fg_1)(f_2 + fg_2)} = \overline{f_1 f_2} .$$

This multiplication satisfies properties (ii), (iii), and (iv) in Definition 14.4. So $F[x]/(f)$ is also a commutative ring, and is called a *quotient* ring of $F[x]$. It is sometimes convenient to describe calculations in $F[x]/(f)$ via congruences mod f:

$$f_1 \equiv f_2 \pmod{f}$$

means that $f_1 = f_2 + fg$ for some $g \in F[x]$, in other words, $\bar{f}_1 = \bar{f}_2$ in $F[x]/(f)$.

Notice that $F[x]/(f)$ is an F-vector space. In fact, if $\deg f = n$, then

$$\dim_F F[x]/(f) = n .$$

Why is this so? Well, any $g \in F[x]$ can be written

$$g = qf + r ,$$

where $\deg r < n$. In other words, $\bar{g} = \bar{r}$ is a linear combination of $1, \bar{x}, \ldots, \bar{x}^{n-1}$. Thus $\{1, \bar{x}, \ldots, \bar{x}^{n-1}\}$ spans $F[x]/(f)$. On the other hand, if for some $a_0, \ldots, a_{n-1} \in F$,

$$a_0 + a_1 \bar{x} + \cdots + a_{n-1}\bar{x}^{n-1} = 0$$

in $F[x]/(f)$, then $f \mid (a_0 + a_1 x + \cdots + a_{n-1}x^{n-1})$, which is not possible. So $\{1, \bar{x}, \ldots, \bar{x}^{n-1}\}$ is linearly independent.

This construction is particularly interesting when f is irreducible. Remember that for p a prime number, $\mathbb{Z}/p\mathbb{Z}$ is a field. The same is true for $F[x]/(f)$ if f is irreducible.

THEOREM 14.5
Let $f \in F[x]$. If f is irreducible, then $F[x]/(f)$ is a field.

PROOF   We must show that every nonzero element in $F[x]/(f)$ has a multiplicative inverse. Suppose $g \in F[x]$ and $f \nmid g$. Then f and g are relatively prime because f is irreducible. So there exist $s, t \in F[x]$ such that

$$1 = sf + tg .$$

Therefore

$$tg \equiv 1 \pmod{f} ,$$

or equivalently, $\bar{t}\bar{g} = 1 \in F[x]/(f)$. Thus $F[x]/(f)$ is a field.   ∎

Examples 14.4

(i) Let $f = x^2 + 1 \in \mathbb{R}[x]$. Since $\deg(x^2 + 1) = 2$,

$$\dim_{\mathbb{R}} \mathbb{R}[x]/(x^2 + 1) = 2 .$$

In fact, we know that

$$\{1, \bar{x}\}$$

is a basis. Since $x^2 + 1$ is irreducible, $\mathbb{R}[x]/(x^2 + 1)$ is a field. Now let $i := \bar{x}$. Then every element can be written in the form $a + bi$, $a, b \in \mathbb{R}$, and

$$i^2 + 1 = \bar{x}^2 + 1 = \overline{x^2 + 1} = 0 .$$

So

$$\mathbb{R}[x]/(x^2 + 1) \cong \mathbb{C} ,$$

the field of complex numbers.

(ii) Let $f(x) = x^2 - 2 \in \mathbb{Q}[x]$. By the Eisenstein criterion, $x^2 - 2$ is irreducible. So $\mathbb{Q}[x]/(x^2 - 2)$ is a field. Since $\deg(x^2 - 2) = 2$,

$$\dim_\mathbb{Q} \mathbb{Q}[x]/(x^2 - 2) = 2 .$$

Define a homomorphism $\epsilon_{\sqrt{2}} : \mathbb{Q}[x] \to \mathbb{R}$ by

$$\epsilon_{\sqrt{2}} (g) = g(\sqrt{2}) .$$

Given a polynomial $g \in \mathbb{Q}[x]$, we can divide it by $x^2 - 2$:

$$g(x) = q(x) \left(x^2 - 2 \right) + (ax + b) ,$$

for some $a, b \in \mathbb{Q}$. Then

$$\epsilon_{\sqrt{2}} (g) = g(\sqrt{2}) = a\sqrt{2} + b .$$

So the image of $\epsilon_{\sqrt{2}}$ is

$$\mathbb{Q}(\sqrt{2}) = \{a\sqrt{2} + b \mid a, b \in \mathbb{Q}\}$$

(see Exercise 1.19). Now for any $g \in \mathbb{Q}[x]$,

$$\epsilon_{\sqrt{2}} (g(x)(x^2 - 2)) = 0 .$$

Therefore $\epsilon_{\sqrt{2}}$ induces a homomorphism

$$\bar{\epsilon}_{\sqrt{2}} : \mathbb{Q}[x]/(x^2 - 2) \to \mathbb{Q}(\sqrt{2}) \subset \mathbb{R} .$$

This is injective because $\mathbb{Q}[x]/(x^2 - 2)$ is a field. Since it is also surjective, it is in fact an isomorphism.

(iii) Suppose $r \in \mathbb{F}_p$ is not a square. Let $f = x^2 - r \in \mathbb{F}_p[x]$. Then f is irreducible. Therefore $\mathbb{F}_p[x]/(f)$ is a field. Again, since $\deg(x^2 - r) = 2$,

$$\dim_{\mathbb{F}_p} \mathbb{F}_p[x]/(f) = 2 .$$

It is not hard to see that the mapping

$$a + b\bar{x} \mapsto \begin{pmatrix} a & b \\ br & a \end{pmatrix}$$

is an isomorphism from $F_p[x]/(f)$ to the field $\mathbb{F}_{p^2}$ defined in Exercise 1.21.

(iv) Take $f(x) = x^{p-1} + \cdots + x + 1 \in \mathbb{Q}[x]$, where p is a prime number. In Example 14.2, we saw that f is irreducible, so that $\mathbb{Q}[x]/(f)$ is a field. If we set $\omega = e^{2\pi i/p}$, then its roots are

$$\left\{\omega, \ldots, \omega^{p-1}\right\} .$$

Let

$$\mathbb{Q}(\omega) = \left\{ a_0 + a_1\omega + \cdots + a_{p-2}\omega^{p-2} \mid a_0, \ldots, a_{p-2} \in \mathbb{Q} \right\} .$$

Define a homomorphism $\epsilon_\omega : \mathbb{Q}[x] \to \mathbb{C}$ by

$$\epsilon_\omega(g) = g(\omega) .$$

The image is just $\mathbb{Q}(\omega)$, and the induced map

$$\bar{\epsilon}_\omega : \mathbb{Q}[x]/(f) \to \mathbb{C} ,$$

is injective. So $\mathbb{Q}(\omega)$ is a field and

$$\mathbb{Q}[x]/(f) \cong \mathbb{Q}(\omega) . \qquad \square$$

REMARK 14.4   We can generalize the constructions in (ii) and (iv) above, which use evaluation maps. Suppose that E and F are fields, with $F \subset E$. So any polynomial in $F[x]$ can be regarded as being a polynomial in $E[x]$ too. Pick an element $\zeta \in E$. Define a map (evaluation at ζ)

$$\epsilon_\zeta : F[x] \to E$$

by

$$\epsilon_\zeta(f) = f(\zeta) .$$

for $f \in F[x]$. Then ϵ_ζ is a ring homomorphism: for any $f, g \in F[x]$,

$$\epsilon_\zeta(f + g) = (f + g)(\zeta) = f(\zeta) + g(\zeta) = \epsilon_\zeta(f) + \epsilon_\zeta(g)$$
$$\epsilon_\zeta(fg) = (fg)(\zeta) = f(\zeta)g(\zeta) = \epsilon_\zeta(f)\epsilon_\zeta(g)$$
$$\epsilon_\zeta(1) = 1(\zeta) = 1 .$$

Now suppose that $f \in F[x]$ is irreducible of degree n, and that $\zeta \in E$ is a root of f. Any $g \in F[x]$ can be written
$$g = qf + r ,$$

where $\deg r < n$. Therefore

$$\epsilon_\zeta(g) = g(\zeta) = q(\zeta)g(\zeta) + r(\zeta) = r(\zeta) .$$

So the image of ϵ_ζ is

$$F(\zeta) := \left\{ a_0 + a_1\zeta + \cdots + a_{n-1}\zeta^{n-1} \mid a_0, \ldots, a_{n-1} \in F \right\} \subset E .$$

The homomorphism ϵ_ζ induces a homomorphism

$$\bar{\epsilon}_\zeta : F[x]/(f) \to E ,$$

since $(gf)(\zeta) = g(\zeta)f(\zeta) = 0$, for any $g \in F[x]$. As $F[x]/(f)$ is a field, $\bar{\epsilon}_\zeta$ must be injective. So $F(\zeta) \cong F[x]/(f)$ is a field, with $F \subset F(\zeta) \subset E$. We shall make heavy use of such fields in coming chapters. ∎

Just as in the integers, we have the Chinese Remainder Theorem (see Theorem 1.5).

THEOREM 14.6 (Chinese Remainder Theorem)
If $p_1, \ldots, p_m \in F[x]$ are pairwise relatively prime, then the m congruences

$$f \equiv g_i \pmod{p_i}, \ 1 \leq i \leq m,$$

have a unique solution modulo $p_1 \cdots p_m$ for any $g_i \in F[x]$.

PROOF   We prove the theorem by induction on m. If $m = 1$, we are looking at a single congruence

$$f \equiv g_1 \pmod{p_1}$$

with the solution $f = g_1$, which is unique modulo p_1. So suppose that the result holds for $m - 1$ congruences, $m > 1$. We want to show that m congruences

$$f \equiv g_i \pmod{p_i}, \ 1 \leq i \leq m,$$

have a solution. By the induction assumption, the first $m - 1$ of these have a solution $f_{m-1} \in F[x]$ and all other solutions are of the form

$$f_{m-1} + up_1 \cdots p_{m-1},$$

for $u \in F[x]$. The mth congruence then becomes

$$up_1 \cdots p_{m-1} \equiv g_m - f_{m-1} \pmod{p_m},$$

which we want to solve for u. Now given that $p_1, \ldots, p_m$ are pairwise relatively prime, it is easy to check that $(p_1 \cdots p_{m-1}, p_m) = 1$. Therefore there exist $s, t \in F[x]$ such that

$$1 = sp_1 \cdots p_{m-1} + tp_m.$$

Multiplying this equation by $g_m - f_{m-1}$ gives

$$g_m - f_{m-1} = (g_m - f_{m-1})sp_1 \cdots p_{m-1} + (g_m - f_{m-1})tp_m.$$

Thus

$$(g_m - f_{m-1})sp_1 \cdots p_{m-1} \equiv g_m - f_{m-1} \pmod{p_m}.$$

So take $u = (g_m - f_{m-1})s$ and let

$$f = f_{m-1} + (g_m - f_{m-1})sp_1 \cdots p_{m-1}.$$

Then

$$f \equiv f_{m-1} \equiv g_i \pmod{p_i}, \ 1 \leq i \leq m - 1,$$

and

$$f \equiv f_{m-1} + (g_m - f_{m-1}) \equiv g_m \pmod{p_m},$$

which are the m congruences we want to solve. Therefore by the principle of induction, there exists a solution for all m.

If f and g are two solutions, then

$$f - g \equiv 0 \pmod{p_i}, \ 1 \le i \le m.$$

Since $p_1, \ldots, p_m$ are relatively prime, it follows that $p_1 \cdots p_m \mid (f - g)$, in other words,

$$f \equiv g \pmod{p_1 \cdots p_m}. \quad \blacksquare$$

REMARK 14.5   The theorem above can also be interpreted the following way (cf. Example 5.5(ii) and Exercise 5.24). For any $f \in F[x]$, let $\bar{f}$ denote its residue class in $F[x]/(p_1 \cdots p_m)$, and $\bar{f}_j$ its residue class in $F[x]/(p_j)$, for $1 \le j \le m$. Then the map

$$\psi : F[x]/(p_1 \cdots p_m) \to F[x]/(p_1) \times \cdots \times F[x]/(p_m)$$

given by

$$\psi : \bar{f} \mapsto (\bar{f}_1, \ldots, \bar{f}_m)$$

is well-defined and is a ring homomorphism (see Exercise 14.19 for direct products of rings). The Chinese Remainder Theorem says precisely that ψ is an isomorphism. $\blacksquare$

Factoring Polynomials over a Finite Field

Earlier we saw how to build up a list of irreducible polynomials in $\mathbb{F}_p[x]$. This is clearly not a good way to find out whether a given polynomial is irreducible. There is a very effective algorithm, discovered by Berlekamp, which will test for irreducibility. In fact, it is actually an algorithm for factoring polynomials over $\mathbb{F}_p$. To explain it, we need to make some preparations.

First, consider the polynomial $x^p - x \in \mathbb{F}_p[x]$. According to Theorem 10.2, every $a \in \mathbb{F}_p$ is a root. Since $x^p - x$ has exactly p roots, these are all of them. So $x^p - x$ factors

$$x^p - x = x(x - 1) \cdots (x - p + 1).$$

If g is any polynomial in $\mathbb{F}_p[x]$, then we can substitute g for x in the previous equation and obtain

$$g(x)^p - g(x) = g(x)(g(x) - 1) \cdots (g(x) - p + 1). \tag{14.1}$$

Now suppose we have a monic polynomial $f \in \mathbb{F}_p[x]$. We can factor it into irreducibles as in Theorem 14.3:

$$f = q_1 \cdots q_r ,$$

where each $q_i = p_i^{m_i}$ for some monic irreducible polynomial p_i and some $m_i \in \mathbb{N}$, and where $q_1, \ldots, q_r$ are pairwise relatively prime. Our algorithm will determine $q_1, \ldots, q_r$. How can one determine $p_1, \ldots, p_r$ from them? Suppose that $q = s^m$, where s is irreducible. There are two cases to consider: (i) $p \nmid m$; (ii) $p \mid m$.

LEMMA 14.4

(i) *If* $p \nmid m$, *then* $(q, q') \neq 1$ *and* $s = q/(q, q')$.

(ii) *If* $m = pn$, *then* $q(x) = q_1(x^p)$ *for some* $q_1 \in \mathbb{F}_p[x]$.

PROOF   In the first case,

$$q' = ms^{m-1}s' .$$

Since s is irreducible, $(s, s') = 1$ and therefore

$$(q, q') = s^{m-1} .$$

It follows that

$$s = q/(q, q') .$$

Now for any polynomial $s \in \mathbb{F}_p[x]$, putting together Exercise 1.4 and Theorem 10.2, we see that

$$s(x)^p = s(x^p) .$$

So if $q = s^{pn}$, then

$$q(x) = (s(x)^p)^n = s^n(x^p) ,$$

and we can take $q_1 = s^n$ to get the second statement.   ∎

To determine s in the second case, we have to apply the lemma again to q_1.

Suppose then that

$$f = q_1 \cdots q_r \in \mathbb{F}_p[x] ,$$

where $q_1, \ldots, q_r$ are pairwise relatively prime. How can we find the factors q_i? Berlekamp's idea is to consider the congruences

$$g \equiv a_1 \pmod{q_1} \quad \ldots \quad g \equiv a_r \pmod{q_r} , \qquad (14.2)$$

where $a_1, \ldots, a_r \in \mathbb{F}_p$. According to the Chinese Remainder Theorem, there is a unique solution g modulo $q_1 \cdots q_r = f$. From solutions to such congruences, you can find the factors $q_1, \ldots, q_r$. On the other hand, a solution g is a solution of the congruence

$$g^p - g \equiv 0 \pmod{f} . \qquad (14.3)$$

This congruence is easy to solve. We shall now explain this in detail.

Recall that $V := \mathbb{F}_p[x]/(f)$ is a vector space over $\mathbb{F}_p$ of dimension $n = \deg f$. The map

$$\psi : g \mapsto g^p$$

is a linear mapping of V to itself. The set of solutions $W \subset V$ of the congruence (14.3) is just the kernel of $\psi - I$. So if we let A be the matrix of ψ, with respect to the basis $\{1, x, \ldots, x^{n-1}\}$ of V, then $W = \ker(A - I)$. This explains how to solve (14.3).

Now if g is a solution of (14.3), then each q_i divides the right-hand side of Equation (14.1) since $f = q_1 \cdots q_r$. As the terms on the right-hand side are relatively prime, this means that for each i, $1 \le i \le r$, there is an $a_i \in \mathbb{F}_p$ such that $q_i \mid g - a_i$. Therefore g is a solution of the congruences (14.2) with this choice of $a_1, \ldots, a_r$. Conversely, suppose g is a solution of (14.2) for some $a_1, \ldots, a_r$. Each term $g - a_i$ occurs in the right-hand side of Equation (14.1). Therefore $q_1 \cdots q_r = f$ divides the right-hand side, and thus g is a solution of (14.3). So r-tuples $(a_1, \ldots, a_r)$ correspond to solutions of (14.3). This gives us the connection between (14.2) and (14.3). In fact we have a linear map from

$$\mathbb{F}_p^r \to W ,$$

given by

$$(a_1 \ldots a_r) \mapsto g , \tag{14.4}$$

where g is the corresponding solution of (14.2). Since this map is an isomorphism, the dimension of W is r.

Finally we must know how to find the factors $q_1, \ldots, q_r$ from solutions $g \in W$. Looking at (14.2) we see that

$$(g - a_i, f) \ne 1$$

for $1 \le i \le r$. It is not hard to see that

$$f = \prod_{a \in \mathbb{F}_p} (g - a, f) , \tag{14.5}$$

for any $g \in W$ (see Exercise 14.23). If $a \ne b$ are in $\mathbb{F}_p$ and $(g - a, f) \ne 1$ and $(g - b, f) \ne 1$, then these two factors of f will be relatively prime. So if the numbers of the r-tuple $(a_1 \ldots a_r)$ corresponding to g are all distinct, then we have r relatively prime factors of f and are finished. If not, then we must take each factor $\tilde{f}$ we have found and repeat the procedure with it. Let us summarize this in the form of an algorithm:

ALGORITHM 14.1

Suppose that f is a polynomial in $\mathbb{F}_p[x]$. To factor f into irreducible polynomials:

(i) *Solve the congruence*

$$g^p - g \equiv 0 \pmod{f} .$$

Let $g_1 = 1, g_2, \ldots, g_r$ be a basis of the solution space, with $\deg g_k < \deg f$, $1 \le k \le r$.

(ii) *Find all $a \in \mathbb{F}_p$ such that*

$$(g_2 - a, f) \ne 0 .$$

If there are r such numbers a, then you have r relatively prime factors of f and are finished by (14.5).

(iii) *Otherwise, take each such factor $\tilde{f}$ and find all $a \in \mathbb{F}_p$ such that*

$$(g_3 - a, \tilde{f}) \ne 0 .$$

Now

$$\deg(g_3 - a, \tilde{f}) < \deg f .$$

This means that as you continue with $g_4, \ldots, g_r$, the process will terminate, and you will end up with r relatively prime factors of f.

(iv) *Each $q_i = p_i^{m_i}$ where p_i is irreducible. To determine p_i, take the derivative of q_i. If $q_i' \ne 0$, then*

$$p_i = q_i / (q_i, q_i') .$$

If $q_i' = 0$, then $q_i(x) = \tilde{q}_i(x^p)$, for some $\tilde{q}_i \in \mathbb{F}_p[x^p]$. The polynomial $\tilde{q}_i$ is in turn a power of an irreducible polynomial in $\mathbb{F}_p[x^p]$. So we return to the beginning of this step and take its derivative.

Let us calculate an example.

Example 14.5
Set

$$f = 3 + x + 6x^2 + 2x^3 + 4x^4 + 5x^5 + x^6 \in \mathbb{F}_7[x] .$$

We want to find solutions of

$$g^7 - g \equiv 0 \pmod{f} .$$

To do this, we compute the matrix A of

$$g \mapsto g^7$$

in $V := \mathbb{F}_7[x]/(f)$ with respect to the basis $\{1, x, \ldots, x^5\}$. We have

$$x^0 = 1$$
$$x^7 = 1 + 2x + x^2 + 4x^3 + 4x^4$$
$$x^{14} = 3 + 4x + 3x^2 + 3x^3 + 2x^4 + 3x^5$$
$$x^{21} = 3 + 3x + x^2 + x^3 + x^4 + 4x^5$$
$$x^{28} = 3 + 5x^2 + 5x^3 + 2x^4$$
$$x^{35} = 3 + 3x^2 + 5x^4 + 5x^5 .$$

Therefore the matrix is

$$A = \begin{pmatrix} 1 & 1 & 3 & 3 & 3 & 3 \\ 0 & 2 & 4 & 3 & 0 & 0 \\ 0 & 1 & 3 & 1 & 5 & 3 \\ 0 & 4 & 3 & 1 & 5 & 0 \\ 0 & 4 & 2 & 1 & 2 & 5 \\ 0 & 0 & 3 & 4 & 0 & 5 \end{pmatrix} .$$

If we row reduce $A - I$, we see that its kernel has the basis $\{(1, 0, 0, 0, 0, 0),$ $(0, 3, 6, 5, 1, 1)\}$, or equivalently, the solution space of (14.3) has a basis $\{1, 3x + 6x^2 + 5x^3 + x^4 + x^5\}$ and $r = 2$. So taking

$$g = 3x + 6x^2 + 5x^3 + x^4 + x^5 ,$$

we look for solutions of (14.2). We find that

$$(g + 4, f) = 5 + 3x + x^2$$
$$(g + 5, f) = 2 + 6x + 2x^3 + x^4$$

and

$$f = \left(5 + 3x + x^2\right)\left(2 + 6x + 2x^3 + x^4\right) .$$

The quadratic $q_1 = 5 + 3x + x^2 \in \mathbb{F}_7[x]$ is clearly irreducible. But $q_2 = 2 + 6x + 2x^3 + x^4$ is not:

$$(q_2, q_2') = 3 + x + x^2 .$$

This quadratic too is irreducible, and $q_2 = (3 + x + x^2)^2$. So we have found the irreducible factors of f:

$$f = \left(5 + 3x + x^2\right)\left(3 + x + x^2\right)^2 . \quad \square$$

Calculations

Mathematica has a package "Algebra" with a function PolynomialExtended-GCD, which calculates (f, g) for two polynomials f and g, as well as polynomials s

and t such that

$$(f, g) = sf + tg .$$

First load the package

> $In[1] := $ «Algebra`PolynomialExtendedGCD`

If we take $f = x^3 + x^2 + x + 1$, $g = x^4 + x^3 + x + 1 \in \mathbb{Q}[x]$, we obtain

> $In[2] := $ PolynomialExtendedGCD[x^3+x^2+x+1,
>   x^4+x^3+x+1]

$$Out[2] = \left\{ 1 + x, \left\{ \frac{1}{2} - \frac{x}{2} - \frac{x^2}{2}, \frac{1}{2} + \frac{x}{2} \right\} \right\}$$

Thus

$$(x^3 + x^2 + x + 1, \ x^4 + x^3 + x + 1) = x + 1 ,$$

and

$$x + 1 = \left(\frac{1}{2} - \frac{x}{2} - \frac{x^2}{2} \right)(x^3 + x^2 + x + 1) + \left(\frac{1}{2} + \frac{x}{2} \right)(x^4 + x^3 + x + 1) .$$

The function can also calculate in $\mathbb{F}_p$. Here is Example 14.1:

> $In[3] := $ PolynomialExtendedGCD[
>   x^4 + x^3 + x^2 + 3x + 2,
>   x^5 - x^4 - x^3 + 2x^2 - x - 2,
>   Modulus -> 11]

$$Out[3] = \{ 1 + x, \{ 8 + 3x + 6x^2 + 4x^3, 2 + 8x + 7x^2 \} \}$$

Mathematica is very useful for a calculation like Example 14.5. We have

> $In[4] := $ $f = 3 + x + 6x^2 + 2x^3 + 4x^4 + 5x^5 + x^6$

$$Out[4] = 3 + x + 6x^2 + 2x^3 + 4x^4 + 5x^5 + x^6$$

To compute x^7, x^{14}, x^{21}, x^{28}, and x^{35} in terms of the basis $1, x, ..., x^6$ of V, we can use the function PolynomialMod, which reduces a polynomial modulo a natural number n and a polynomial f.

> $In[5] := $ PolynomialMod[x^7, {f,7}]

$$Out[5] = 1 + 2x + x^2 + 4x^3 + 4x^4$$

In[6]:= PolynomialMod[x^14, {f,7}]

Out[6]= $3 + 4x + 3x^2 + 3x^3 + 2x^4 + 3x^5$

In[7]:= PolynomialMod[x^21, {f,7}]

Out[7]= $3 + 3x + x^2 + x^3 + x^4 + 4x^5$

In[8]:= PolynomialMod[x^28, {f,7}]

Out[8]= $3 + 5x^2 + 5x^3 + 2x^4$

In[9]:= PolynomialMod[x^35, {f,7}]

Out[9]= $3 + 3x^2 + 5x^4 + 5x^5$

Then

In[10]:= A = Transpose[{{1,0,0,0,0,0},{1,2,1,4,4,0},
{3,4,3,3,2,3}, {3,3,1,1,1,4},
{3,0,5,5,2,0},{3,0,3,0,5,5}}];

In[11]:= MatrixForm[A]

Out[11]=
$$\begin{pmatrix} 1 & 1 & 3 & 3 & 3 & 3 \\ 0 & 2 & 4 & 3 & 0 & 0 \\ 0 & 1 & 3 & 1 & 5 & 3 \\ 0 & 4 & 3 & 1 & 5 & 0 \\ 0 & 4 & 2 & 1 & 2 & 5 \\ 0 & 0 & 3 & 4 & 0 & 5 \end{pmatrix}$$

To compute the kernel of A- I, we can use the function Nullspace.

In[12]:= NullSpace[A-IdentityMatrix[6], Modulus->7]

Out[12]= {{0,3,6,5,1,1},{1,0,0,0,0,0}}

Now we set

In[13]:= g = 3 x + 6 x^2 + 5 x^3 + x^4 + x^5

Out [13] = $3x + 6x^2 + 5x^3 + x^4 + x^5$

To check the greatest common divisors $(g + a, f)$ for $a \in \mathbb{F}_7$, we use the function `PolynomialGCD`:

In [14] := `PolynomialGCD [ g + 4, f, Modulus->7 ]`

Out [14] = $5 + 3x + x^2$

In [15] := `PolynomialGCD [ g + 5, f, Modulus->7 ]`

Out [15] = $2 + 6x + 2x^3 + x^4$

We check the greatest common divisor of this quartic with its derivative:

In [16] := `PolynomialGCD [ 2 + 6x + 2x^3 + x^4,`
            `PolynomialMod [`
             `D[2 + 6x + 2x^3 + x^4, x], 7 ],`
            `Modulus->7 ]`

Out [16] = $3 + x + x^2$

Lastly, we verify the factorization.

In [17] := `PolynomialMod [ Expand[(5 + 3 x + x^2)*`
                  `(3 + x + x^2)^2], 7 ]`

Out [17] = $3 + x + 6x^2 + 2x^3 + 4x^4 + 5x^5 + x^6$

Mathematica does have a built-in function that implements Berlekamp's algorithm:

In [18] := `Factor[ x^6 + 5 x^5 + 4 x^4 + 2 x^3 + 6 x^2`
                  `+ x + 3, Modulus -> 7 ]`

Out [18] = $(3 + x + x^2)^2 \ (5 + 3x + x^2)$

We can use this to check that $x^5 - 5x + 12$ is irreducible modulo 7:

In [19] := `Factor[ x^5 - 5 x + 12, Modulus -> 7 ]`

Out [19] = $5 + 2x + x^5$

You can also factor polynomials over $\mathbb{Q}$:

In [20] := `Factor[x^6 + x^5 + 4 x^4 + 2 x^3 + 6 x^2 + x + 3]`

Out [20] = $(1 + x^2) \ (3 + x + 3x^2 + x^3 + x^4)$

Exercises

14.1 Let $p \in F[x]$ be irreducible, and suppose $p \mid (f_1 \cdots f_r)$, where $f_1, \ldots, f_r \in F[x]$. Prove that $p \mid f_i$ for some i, $1 \le i \le r$.

14.2 Suppose that $p_1, \ldots, p_m$, $m > 2$, are pairwise relatively prime. Prove that $(p_1 \cdots p_{m-1})$ and p_m are relatively prime.

14.3 (a) Let $f(x) = a_n x^n + \cdots + a_1 x + a_0$ be a polynomial with integer coefficients. Show that if a rational number a/b, with $(a, b) = 1$, is a root of f, then $a \mid a_0$ and $b \mid a_n$.

   (b) Find the rational roots of $8 - 38x + 27x^2 + 47x^3 - 11x^4 + 15x^5$.

14.4 Show that a polynomial in $F[x]$ of degree 2 or 3 is reducible if and only if it has a root.

14.5 Make a list of the monic irreducible polynomials of degree less than 4 in $\mathbb{F}_3[x]$.

14.6 Write a *Mathematica* function that lists the monic irreducible polynomials of degree less than n in $\mathbb{F}_p[x]$ by making a sieve.

14.7 Decide whether the following polynomials are reducible or irreducible:

   (a) $x^5 + x^3 + x^2 + x + 1 \in \mathbb{F}_2[x]$;

   (b) $x^4 + 2x^2 + x + 2 \in \mathbb{F}_3[x]$;

   (c) $x^4 + x^2 + 1 \in \mathbb{F}_3[x]$;

   (d) $x^5 + 6x^4 - 54x^3 + 12x^2 + 72x + 24 \in \mathbb{Q}[x]$;

   (e) $x^4 - 10x^2 + 1 \in \mathbb{Q}[x]$;

   (f) $x^7 + 3x + 5 \in \mathbb{Q}[x]$.

14.8 • Let f be a monic polynomial with integer coefficients. Suppose that $g \in \mathbb{Q}[x]$ is monic and divides f. Prove that the coefficients of g are integers too.

14.9 Let p be prime. By factoring $x^{p-1} - 1 \in \mathbb{F}_p[x]$, show that

$$(p - 1)! \equiv -1 \pmod{p}.$$

14.10 • Let F be a field, and $K \subset F$, its prime field. Show that if α is an automorphism of F, then α fixes K, in other words $\alpha(a) = a$ for all $a \in K$.

14.11 (a) Show that the set of automorphisms of a field forms a group under composition.

(b) Compute the group of automorphisms of

    (i) $\mathbb{Q}(\sqrt{2})$,

    (ii) $\mathbb{Q}(\omega)$, where $\omega = e^{2\pi i/5}$,

    (iii) $\mathbb{Q}(\omega)$, where $\omega = e^{\pi i/4}$.

14.12 • Let F_1 and F_2 be fields, and $\psi : F_1 \to F_2$ an isomorphism. Then ψ induces a mapping $\psi_* : F_1[x] \to F_2[x]$ as follows. For

$$f(x) = a_n x^n + \cdots + a_1 x + a_0 \,,$$

with $a_n, \ldots, a_1, a_0 \in F_1$, let

$$b_j = \psi(a_j), \quad 0 \le j \le n \,,$$

and then set

$$\psi_*(f)(x) = b_n x^n + \cdots + b_1 x + b_0 \,.$$

Prove that ψ_* is an isomorphism of rings.

14.13 Suppose $f \in F[x]$ is reducible. Prove that $F[x]/(f)$ is not a field (cf. Exercise 1.18).

14.14 Prove that the mapping in Example 14.4(iii) is an isomorphism of fields.

14.15 • Let F be a finite field of characteristic p with q elements.

(a) Verify that F is a vector space over $\mathbb{F}_p$. If $\dim_{\mathbb{F}_p} F = n$, show that $q = p^n$.

(b) Suppose $f \in F[x]$ is an irreducible polynomial of degree n. Prove that $E := F[x]/(f)$ is a finite field. How many elements does it have?

(c) Construct fields with

    (i) 8 elements;

    (ii) 9 elements;

    (iii) 125 elements.

14.16 Let G be a finite abelian group.

(a) Given $\alpha, \beta \in G$ show that there exists an element in G of order

$$\mathrm{lcm}(|\alpha|, |\beta|) \,.$$

Suggestion: use Exercise 5.8.

(b) Let

$$r = \max_{\alpha \in G} |\alpha| \,.$$

Show that $|\alpha| \mid r$ for all $\alpha \in G$.

(c) Prove that the multiplicative group of a finite field is cyclic.

14.17 Let R be a commutative ring and let $I \subset R$ be a subgroup such that

$$rI \subset I$$

for all $r \in R$. Prove that there is a well-defined multiplication on the quotient group R/I induced by the multiplication on R, and that with this multiplication, R/I is a ring. Such a subgroup I is called an *ideal* of R, and the ring R/I is called the quotient ring of I. Let S be another ring and $\psi : R \to S$ a ring homomorphism. Show that ker ψ is an ideal. Prove that if ψ is surjective, then it induces an isomorphism of rings:

$$R/I \cong S .$$

14.18 Prove that every ideal I of $F[x]$ is of the form $I = (f)$ for some $f \in F[x]$.

14.19 Let R and S be two rings. For any $r_1, r_2 \in R$, $s_1, s_2 \in S$, define

(a) $(r_1, s_1) + (r_2, s_2) := (r_1 + r_2, s_1 + s_2)$.

(b) $(r_1, s_1)(r_2, s_2) := (r_1 r_2, s_1 s_2)$.

Show that with these operations, $R \times S$ is a ring with multiplicative identity $(1_R, 1_S)$.

14.20 • Prove that a polynomial $f \in F[x]$ has a repeated factor if and only if $(f, f') \neq 1$.

14.21 • Prove that for any $g \in \mathbb{F}_p[x]$,

$$g(x)^p = g(x^p) .$$

14.22 • Let F be a field of characteristic p. Show that the map $\psi : F \to F$, given by

$$\psi(a) = a^p ,$$

is a homomorphism. Prove that if F is finite, then ψ is an automorphism of F, in particular that every element of F is a pth power.

14.23 (a) Let F be a field and $f \in F[x]$. Suppose that $g, h \in F[x]$ are relatively prime. Show that

$$(f, gh) = (f, g)(f, h) .$$

(b) Prove that Equation (14.5) holds.

14.24 Prove that the Mapping (14.4) is linear, and is an isomorphism: $\mathbb{F}_p^r \to W$.

14.25 Use Berlekamp's algorithm to show that $x^5 - 5x + 12 \in \mathbb{F}_7[x]$ is irreducible.

14.26 Use Berlekamp's algorithm to factor $x^6 + 3x^5 + x^4 + x^3 + 5x^2 + x + 4 \in \mathbb{F}_7[x]$.

14.27 Use Berlekamp's algorithm to factor
$x^8 + x^6 + 10x^4 + 10x^3 + 8x^2 + 2x + 8 \in \mathbb{F}_{13}[x]$.

Chapter 15

Symmetric Polynomials

The coefficients of a polynomial in one variable are symmetric functions of its roots. So are other quantities, like the discriminant of the polynomial. In this chapter, we will discuss symmetric polynomials and their basic properties. First, we briefly discuss polynomials in more than one variable.

Polynomials in Several Variables

Let F be a field. A polynomial f in n variables $x_1, x_2, \ldots, x_n$, with coefficients in F, is a finite sum

$$f(x_1, \ldots, x_n) = \sum_{i_1, \ldots, i_n} a_{i_1 \cdots i_n} x_1^{i_1} \cdots x_n^{i_n} \, ,$$

where the coefficients $a_{i_1 \cdots i_n}$ lie in F. We denote by $F[x_1, \ldots, x_n]$, the set of all polynomials in $x_1, \ldots, x_n$ with coefficients in F. The degree of a monomial $x_1^{i_1} \cdots x_n^{i_n}$ is $i_1 + \cdots + i_n$. The *degree* of a polynomial f is the largest degree of a monomial with a nonzero coefficient in f. For example, the degree of

$$x_1^2 x_2^3 x_3 + x_1 x_3^4 \in F[x_1, x_2, x_3] \, ,$$

is 6. You can add and multiply two such polynomials in the obvious way. With these two operations, $F[x_1, \ldots, x_n]$ becomes a commutative ring. The zero element is the polynomial, all of whose coefficients are 0. The constant polynomials form a subring isomorphic to the field of coefficients F.

Just as with one variable, we can define rational functions in n variables. A *rational function* in $x_1, \ldots, x_n$ over F is a quotient f/g, where $f, g \in F[x_1, \ldots, x_n]$, $g \neq 0$. We identify f/g with kf/kg for any $k \in F[x_1, \ldots, x_n]$. The set of rational functions in $x_1, \ldots, x_n$ over F is denoted by $F(x_1, \ldots, x_n)$. If you define addition and multiplication in the same way as for one variable, then $F(x_1, \ldots, x_n)$ becomes a field, called the field of rational functions in $x_1, \ldots, x_n$.

Symmetric Polynomials and Functions

A symmetric polynomial is one that is symmetric in the variables $x_1, \ldots, x_n$:

DEFINITION 15.1   *A polynomial $f \in F[x_1, \ldots, x_n]$ is symmetric if*

$$f\left(x_{\alpha(1)}, \ldots, x_{\alpha(n)}\right) = f(x_1, \ldots, x_n) ,$$

for all $\alpha \in S_n$.

For example,

$$x_1^2 x_2 + x_1^2 x_3 + x_1 x_2^2 + x_2^2 x_3 + x_1 x_3^2 + x_2 x_3^2 \in F[x_1, x_2, x_3]$$

is symmetric. It is easy to see that if f_1 and f_2 are symmetric, then so are $f_1 + f_2$ and $f_1 f_2$. It follows that the set of all symmetric polynomials in $F[x_1, \ldots, x_n]$ is a subring of $F[x_1, \ldots, x_n]$. A rational function $h \in F(x_1, \ldots, x_n)$ is called symmetric if

$$h\left(x_{\alpha(1)}, \ldots, x_{\alpha(1)}\right) = h(x_1, \ldots, x_n) ,$$

for all $\alpha \in S_n$. The set of all symmetric functions in $F(x_1, \ldots, x_n)$ is a field.

Now suppose that

$$f(x) = x^n + a_{n-1} x^{n-1} + \cdots + a_0 \in F[x] ,$$

and that f has n roots $\zeta_1, \ldots, \zeta_n \in F$. So

$$f(x) = (x - \zeta_1) \cdots (x - \zeta_n) .$$

Expanding the product, you obtain formulas for the coefficients $a_0, \ldots, a_{n-1}$:

$$a_0 = (-1)^n \zeta_1 \cdots \zeta_n$$

$$\vdots \quad \vdots$$

$$a_{n-j} = (-1)^j \sum_{i_1 < \cdots < i_j} \zeta_{i_1} \cdots \zeta_{i_j}$$

$$\vdots \quad \vdots$$

$$a_{n-1} = -\sum_{j=1}^{n} \zeta_j .$$

These expressions are symmetric in $\zeta_1, \ldots, \zeta_n$. The corresponding symmetric polynomials are called the elementary symmetric polynomials:

DEFINITION 15.2   *The jth elementary symmetric polynomial in $x_1, \ldots, x_n$ is*

$$s_j(x_1, \ldots, x_n) := \sum_{i_1 < \cdots < i_j} x_{i_1} \cdots x_{i_j} \,,$$

for $1 \le j \le n$.

Thus

$$a_{n-j} = (-1)^j s_j(\zeta_1, \ldots, \zeta_n) \,.$$

The second reason that elementary symmetric polynomials are important is that every symmetric polynomial can be expressed in terms of them. The proof actually gives an algorithm for doing so. Let us see how it works in an example before looking at the general case. Take

$$h(x_1, x_2, x_3) = x_1^3 + x_2^3 + x_3^3 \,.$$

We will write h in terms of

$$s_1 = x_1 + x_2 + x_3$$
$$s_2 = x_1 x_2 + x_1 x_3 + x_2 x_3$$
$$s_3 = x_1 x_2 x_3 \,.$$

To begin with, notice that the only way to obtain x_1^3 using s_1, s_2, and s_3, is to take s_1^3. Then

$$h = s_1^3 - 3\left(x_1^2 x_2 + x_1^2 x_3 + x_2^2 x_3 + x_1 x_2^2 + x_1 x_3^2 + x_2 x_3^2\right) - 6x_1 x_2 x_3 \,.$$

Now the first term in the difference $h - s_1^3$ is $x_1^2 x_2$. This occurs only in $s_1 s_2$, and

$$h = s_1^3 - 3s_1 s_2 + 3x_1 x_2 x_3 \,.$$

Of course what remains on the right-hand side is just $3s_3$. So we end up with

$$h = s_1^3 - 3s_1 s_2 + 3s_3 \,.$$

To carry out this procedure for an arbitrary symmetric polynomial, we will have to argue by induction. For this, we must order the monomials. The simplest way to do so is to order them *lexicographically*. Suppose that we have two monomials, $x_1^{i_1} \cdots x_n^{i_n}$ and $x_1^{j_1} \cdots x_n^{j_n}$. Then we define

$$x_1^{i_1} \cdots x_n^{i_n} \succ x_1^{j_1} \cdots x_n^{j_n} \,,$$

if

$$i_1 = j_1, \ldots, i_r = j_r, \text{ but } i_{r+1} > i_{r+1} \,,$$

for some r, $0 \le r < n$. So for example,

$$x_1^3 \succ x_1^2 x_2 \succ x_1^2 x_3 \succ x_1 x_2^2 \,.$$

This ordering has the property that, for any two monomials m_1 and m_2,

$$m_1 \succ m_2, \qquad m_1 = m_2 \quad \text{or} \quad m_2 \succ m_1.$$

Notice that the greatest monomial in s_j is $x_1 \cdots x_j$. Therefore in an expression

$$s_1^{j_1} s_2^{j_2} \cdots s_n^{j_n},$$

the greatest monomial will be

$$x_1^{j_1} \left(x_1^{j_2} x_2^{j_2} \right) \cdots \left(x_1^{j_n} \cdots x_n^{j_n} \right) = x_1^{j_1 + j_2 + \cdots + j_n} x_2^{j_2 + \cdots + j_n} \cdots x_n^{j_n}. \qquad (15.1)$$

Now let f be a symmetric polynomial in $x_1, \ldots, x_n$. Order its monomials lexicographically. Suppose that $x_1^{i_1} \cdots x_n^{i_n}$ is the greatest monomial that occurs in f. To write it in terms of the elementary symmetric polynomials, we compare its exponents with those of the monomial (15.1). This gives us the system of linear equations:

$$
\begin{aligned}
i_1 &= j_1 + j_2 + \cdots + j_n \\
i_2 &= j_2 + \cdots + j_n \\
&\ \ \vdots \qquad\qquad \ddots \\
i_n &= j_n.
\end{aligned}
$$

This system is easy to solve and has the unique solution:

$$
\begin{aligned}
j_1 &= i_1 - i_2 \\
j_2 &= i_2 - i_3 \\
&\ \ \vdots \qquad\quad \ddots \\
j_n &= i_n.
\end{aligned}
$$

Therefore we must use $s_1^{i_1 - i_2} s_2^{i_2 - i_3} \cdots s_n^{i_n}$ to get the term $x_1^{i_1} \cdots x_n^{i_n}$. The difference

$$h - s_1^{i_1 - i_2} s_2^{i_2 - i_3} \cdots s_n^{i_n},$$

is again a symmetric polynomial and only has lesser monomials in it. So arguing by induction, it can be written in a unique way in terms of $s_1, \ldots, s_n$. This gives us the theorem:

THEOREM 15.1 (Fundamental Theorem of Symmetric Polynomials)
A symmetric polynomial $f \in F[x_1, \ldots, x_n]$ can be written in a unique way as a polynomial in the elementary symmetric polynomials.

COROLLARY 15.1
There are no algebraic relations among the elementary symmetric polynomials.

PROOF By an algebraic relation among $s_1, \ldots, s_n$ is meant a polynomial $g \neq 0$ in n variables such that $g(s_1, \ldots, s_n) = 0$. If there were such a polynomial relation among them, then there would be more than one way to express the symmetric polynomial 0 in terms of them. ∎

Examples 15.1

(i) Let

$$f(x) = x^3 + ax^2 + bx + c \in F[x],$$

with roots ζ_1, ζ_2 and ζ_3 in F. Thus

$$
\begin{aligned}
a &= -s_1(\zeta_1, \zeta_2, \zeta_3) = -\zeta_1 - \zeta_2 - \zeta_3 \\
b &= s_2(\zeta_1, \zeta_2, \zeta_3) = \zeta_2\zeta_3 + \zeta_1\zeta_3 + \zeta_1\zeta_2 \\
c &= -s_3(\zeta_1, \zeta_2, \zeta_3) = -\zeta_1\zeta_2\zeta_3 \,.
\end{aligned}
$$

Now we are going to see how to write the monic cubic $g \in F[x]$ whose roots are ζ_1^2, ζ_2^2, and ζ_3^2, in terms of a, b, and c. To do this we must write $s_1(\zeta_1^2, \zeta_2^2, \zeta_3^2)$, $s_2(\zeta_1^2, \zeta_2^2, \zeta_3^2)$, and $s_3(\zeta_1^2, \zeta_2^2, \zeta_3^2)$ in terms of a, b, and c. We have

$$
\begin{aligned}
s_1\left(\zeta_1^2, \zeta_2^2, \zeta_3^2\right) &= \zeta_1^2 + \zeta_2^2 + \zeta_3^2 \\
&= (\zeta_1 + \zeta_2 + \zeta_3)^2 - 2(\zeta_2\zeta_3 + \zeta_1\zeta_3 + \zeta_1\zeta_2) \\
&= a^2 - 2b
\end{aligned}
$$

$$
\begin{aligned}
s_2\left(\zeta_1^2, \zeta_2^2, \zeta_3^2\right) &= \zeta_2^2\zeta_3^2 + \zeta_1^2\zeta_3^2 + \zeta_1^2\zeta_2^2 \\
&= (\zeta_2\zeta_3 + \zeta_1\zeta_3 + \zeta_1\zeta_2)^2 - 2\left(\zeta_1^2\zeta_2\zeta_3 + \zeta_1\zeta_2^2\zeta_3 + \zeta_1\zeta_2\zeta_3^2\right) \\
&= (\zeta_2\zeta_3 + \zeta_1\zeta_3 + \zeta_1\zeta_2)^2 - 2(\zeta_1\zeta_2\zeta_3)(\zeta_1 + \zeta_2 + \zeta_3) \\
&= b^2 - 2ac
\end{aligned}
$$

$$
\begin{aligned}
s_3\left(\zeta_1^2, \zeta_2^2, \zeta_3^2\right) &= \zeta_1^2\zeta_2^2\zeta_3^2 \\
&= c^2 \,.
\end{aligned}
$$

Therefore the cubic g is given by

$$
\begin{aligned}
g(x) &= x^3 - s_1\left(\zeta_1^2, \zeta_2^2, \zeta_3^2\right)x^2 + s_2\left(\zeta_1^2, \zeta_2^2, \zeta_3^2\right)x - s_3\left(\zeta_1^2, \zeta_2^2, \zeta_3^2\right) \\
&= x^3 - \left(a^2 - 2b\right)x^2 + \left(b^2 - 2ac\right)x - c^2 \,.
\end{aligned}
$$

(ii) Suppose we have a polynomial

$$f(x) = x^n + a_{n-1}x^{n-1} + \cdots + a_0 \in F[x],$$

with roots $\zeta_1, \ldots, \zeta_n \in F$, none of which are 0. Then we can write a monic polynomial $g \in F[x]$ whose roots are the reciprocals $1/\zeta_1, \ldots, 1/\zeta_n$, in terms of $a_0, \ldots, a_{n-1}$. What we must do is to express $s_k(1/\zeta_1, \ldots, 1/\zeta_n)$, $0 \le k < n$, in terms of $a_0, \ldots, a_{n-1}$. Well,

$$s_k(1/\zeta_1, \ldots, 1/\zeta_n) = \sum_{i_1 < \cdots < i_k} \frac{1}{\zeta_{i_1} \cdots \zeta_{i_k}} .$$

If we multiply this by $(\zeta_1 \cdots \zeta_n)$, we obtain

$$(\zeta_1 \cdots \zeta_n) \, s_k(1/\zeta_1, \ldots, 1/\zeta_n) = \sum_{j_1 < \cdots < j_{n-k}} \zeta_{j_1} \cdots \zeta_{j_{n-k}} .$$

Therefore

$$s_k(1/\zeta_1, \ldots, 1/\zeta_n) = s_{n-k}(\zeta_1, \ldots, \zeta_n)/s_n(\zeta_1, \ldots, \zeta_n)$$
$$= (-1)^k a_k/a_0 .$$

It follows that

$$g(x) = x^n + \cdots + (a_{n-k}/a_0) \, x^k + \cdots + (1/a_0) .$$

Now that we have the answer, it is clear that we could get it more easily:

$$x^n f\left(\frac{1}{x}\right) = 1 + \cdots + a_{n-k} x^k + \cdots + a_0 x^n = a_0 g(x) ,$$

and the roots of $x^n f(1/x)$ are $1/\zeta_1, \ldots, 1/\zeta_n$. $\square$

Sums of Powers

There is another simple type of symmetric polynomial that occurs frequently, namely the sums of the jth powers. Let

$$p_j(x_1, \ldots, x_n) := \sum_{i=1}^{n} x_i^j .$$

By the Fundamental Theorem there must be formulas that express $p_1, \ldots, p_n$ in terms of $s_1, \ldots, s_n$. These can be obtained recursively from *Newton's identities*.

THEOREM 15.2

$$0 = p_1 - s_1$$
$$0 = p_2 - s_1 p_1 + 2s_2$$
$$0 = p_3 - s_1 p_2 + s_2 p_1 - 3s_3$$
$$\vdots$$
$$0 = p_n - s_1 p_{n-1} + s_2 p_{n-2} + \cdots + (-1)^n n s_n .$$

PROOF   We have for $n \geq r > 1$,

$$p_r = \sum_j x_j^r$$

$$s_1 p_{r-1} = \sum_j x_j \sum_k x_k^{r-1} = \sum_j x_j^r + \sum_{j_1 \neq j_2} x_{j_1} x_{j_2}^{r-1}$$

$$s_2 p_{r-2} = \sum_{j_1 < j_2} x_{j_1} x_{j_2} \sum_k x_k^{r-2} = \sum_{j_1 \neq j_2} x_{j_1} x_{j_2}^{r-1} + \sum_{j_1 < j_2 < j_3} \sum_i x_{j_1} x_{j_i}^{r-2} x_{j_3}$$

$$\vdots \qquad\qquad\qquad \vdots$$

$$s_{r-1} p_1 = \sum_{j_1 < \cdots < j_{r-1}} x_{j_1} \cdots x_{j_{r-1}} \sum_k x_k = \sum_{j_1 < \cdots < j_{r-1}} \sum_i x_{j_1} \cdots x_{j_i}^2 \cdots x_{j_{r-1}}$$
$$+ r \sum_{j_1 < \cdots < j_r} x_{j_1} \cdots x_{j_r}$$

$$r s_r = r \sum_{j_1 < \cdots < j_r} x_{j_1} \cdots x_{j_r} .$$

Taking the alternating sum of these equations gives us the formula.   ∎

Notice that if $\operatorname{char} F = 0$, then these identities also allow us to write $s_1, \ldots, s_n$ in terms of $p_1, \ldots, p_n$.

Discriminants

Let F be a field of characteristic $\neq 2$ and $f(x) = x^2 + a_1 x + a_0 \in F[x]$, a quadratic polynomial with roots ζ_1 and ζ_2 in F. By completing the square, we get the well-known formula:

$$\zeta_1 = \frac{-a_1 + \sqrt{\Delta}}{2} , \qquad \zeta_2 = \frac{-a_1 - \sqrt{\Delta}}{2} ,$$

where $\Delta = a_1^2 - 4a_0$ is the discriminant of f. This formula tells us that

$$\Delta = (\zeta_1 - \zeta_2)^2 \; .$$

Now suppose that $f \in F[x]$ is a polynomial of degree n with roots $\zeta_1, \ldots, \zeta_n \in F$. By analogy,

$$\Delta := \prod_{i<j} (\zeta_i - \zeta_j)^2 = (-1)^{n(n-1)/2} \prod_{i \neq j} (\zeta_i - \zeta_j)$$

is called the *discriminant* of f. The most obvious property of Δ is that it is nonzero if and only if the roots of f are distinct. In the chapters to come, we shall see some of its more subtle properties.

Interchanging two roots ζ_i and ζ_j does not change Δ. Since S_n is generated by the set of all transpositions, Δ is therefore symmetric in $\zeta_1, \ldots, \zeta_n$. So define a polynomial

$$\Delta (x_1, \ldots, x_n) := \prod_{i<j} (x_i - x_j)^2 \in F[x_1, \ldots, x_n] \; .$$

Then Δ is symmetric in $x_1, \ldots, x_n$. Using the algorithm above, you can compute that

$$\Delta (x_1, x_2, x_3) = s_1^2 s_2^2 - 4s_2^3 - 4s_1^3 s_3 + 18 s_1 s_2 s_3 - 27 s_3^2 \; .$$

With software it is easy to calculate $\Delta(x_1, \ldots, x_n)$ in terms of the elementary symmetric polynomials for larger n. Notice that the polynomial

$$\delta (x_1, \ldots, x_n) := \prod_{i<j} (x_i - x_j)$$

is not symmetric. If you interchange two variables you get $-\delta(x_1, \ldots, x_n)$ (see Exercise 15.11).

Software

Mathematica has a package "SymmetricPolynomials.m" with a function `SymmetricReduction` that implements the algorithm in Theorem 15.1. It requires that you list the variables in your symmetric polynomial, and name the elementary symmetric polynomials. First load the package:

```
In[1]:= « Algebra`SymmetricPolynomials`
```

Here is p_3 in terms of s_1, s_2, and s_3:

```
In[2]:= SymmetricReduction[ u^3 + v^3 + w^3,
              {u,v,w}, {s1,s2,s3} ]
```

Out [2] = {s1^3 $-$ 3 s1 s2 + 3 s3, 0}

Actually the function will take an arbitrary polynomial, and write it as symmetric polynomial and a remainder. This is where the second term 0 comes from. If you are only interested in symmetric polynomials, you can get rid of the zero remainder by applying the function First:

In [3] := First [%]

Out [3] = s1^3 $-$ 3 s1 s2 + 3 s3

Here is the discriminant Delta in four variables x1, x2, x3, x4, written in terms of the elementary symmetric polynomials:

In [4] := Delta = ((x1-x2)(x2-x3)(x3-x4)(x1-x3)
            (x2-x4)(x1-x4))^2

Out [4] = $(x1 - x2)^2 (x2 - x3)^2 (x3 - x4)^2 (x1 - x3)^2 (x2 - x4)^2 (x1 - x4)^2$

In [5] := First [SymmetricReduction [Expand [Delta],
            {x1,x2,x3,x4}, {s1,s2,s3,s4}]]

Out [5] = s1^2 s2^2 s3^2 $-$ 4 s2^3 s3^2 $-$ 4 s1^3 s3^3 + 18 s1 s2 s3^3
            $-$ 27 s3^4 $-$ 4 s1^2 s2^3 s4 + 16 s2^4 s4 + 18 s1^3 s2 s3 s4
            $-$ 80 s1 s2^2 s3 s4 $-$ 6 s1^2 s3^2 s4 + 144 s2 s3^2 s4
            $-$ 27 s1^4 s4^2 + 144 s1^2 s2 s4^2 $-$ 128 s2^2 s4^2
            $-$ 192 s1 s3 s4^2 + 256 s4^3

Exercises

15.1 Prove that the set of all symmetric polynomials in $F[x_1, \ldots, x_n]$ is a subring of $F[x_1, \ldots, x_n]$ and that the set of all symmetric functions in $F(x_1, \ldots, x_n)$ is a field.

15.2 Write $x_1^5 + x_2^5$ in terms of s_1 and s_2.

15.3 Suppose that

$$f(x) = x^3 + a_2 x^2 + a_1 x + a_0 \in F[x]$$

has roots $\zeta_1, \zeta_2, \zeta_3 \in F$. Write down the polynomial with roots $\zeta_2 \zeta_3, \zeta_1 \zeta_3$, and $\zeta_1 \zeta_2$ in terms of the coefficients of f.

15.4   (a) Give an algorithm that expresses an arbitrary symmetric polynomial in $\mathbb{Q}[x_1, \ldots, x_n]$ in terms of $p_1, \ldots, p_n$. Do not go via the elementary symmetric polynomials.

(b) Write $\Delta(x_1, x_2, x_3)$ in terms of p_1, p_2, and p_3.

15.5   Let $f \in \mathbb{Q}[x]$ be a cubic. Suppose that the sum of its roots is 0, the sum of the squares of its roots is 1, and the sum of the cubes of its roots is 2. What is the sum of the fourth powers of its roots?

15.6   • Let
$$f(x) = x^4 + b_2 x^2 + b_1 x + b_0 \in F[x] ,$$
where $\operatorname{char} F$ is not 2, be a quartic with roots $\zeta_1, \zeta_2, \zeta_3, \zeta_4 \in F$. Set
$$\eta_1 = (\zeta_1 + \zeta_2)(\zeta_3 + \zeta_4)$$
$$\eta_2 = (\zeta_1 + \zeta_3)(\zeta_2 + \zeta_4)$$
$$\eta_3 = (\zeta_1 + \zeta_4)(\zeta_2 + \zeta_3) ,$$

and let
$$r(x) = (x - \eta_1)(x - \eta_2)(x - \eta_3) .$$
Prove that
$$r(x) = x^3 - 2b_2 x^2 + \left(b_2^2 - 4b_0\right) x + b_1^2 .$$

15.7   Derive Newton's identities for p_{n+r}, $r > 0$: show that
$$p_{n+r} - s_1 p_{n+r-1} + s_2 p_{n+r-2} + \cdots + (-1)^n s_n p_r = 0 .$$

15.8   Prove that there exist polynomials $r_n \in \mathbb{Q}[x]$ such that
$$x^n + \frac{1}{x^n} = r_n\left(x + \frac{1}{x}\right) ,$$
for $n \geq 1$. Suggestion: derive a recursion formula for $x^n + 1/x^n$ from Newton's identities.

15.9   Let F be a field of characteristic 0. A polynomial $f \in F[x]$ of degree n is called a *reciprocal polynomial* if
$$x^n f\left(\frac{1}{x}\right) = f(x) .$$

(a) If f is a reciprocal polynomial and $\zeta \in F$, $\zeta \neq 0$ is a root of f, verify that $1/\zeta$ is also a root of f.

(b) Suppose that n is odd and that f is a reciprocal polynomial. Show that -1 is a root of f, and that
$$g(x) := f(x)/(x + 1)$$
is also a reciprocal polynomial.

(c) Suppose that n is even, and

$$f(x) = a_n x^n + \cdots + a_1 x + a_0 .$$

Show that f is a reciprocal polynomial if and only if

$$a_{n-k} = a_k ,$$

for all k.

(d) Let f be a reciprocal polynomial, and let $\zeta \in F$ be a root of f. Prove that $\zeta \neq 0$.

(e) If $n = 2m$ and f is a reciprocal polynomial, prove that

$$\frac{1}{x^m} f(x) = g \left(x + \frac{1}{x} \right)$$

for some polynomial $g \in F[x]$ of degree m.

15.10 (a) Show that

$$(-1)^{\frac{n(n-1)}{2}} \prod_{i<j} (x_i - x_j) = \begin{vmatrix} 1 & x_1 & x_1^2 & \cdots & x_1^{n-1} \\ 1 & x_2 & x_2^2 & \cdots & x_2^{n-1} \\ \vdots & \vdots & \vdots & & \vdots \\ 1 & x_n & x_n^2 & \cdots & x_n^{n-1} \end{vmatrix} .$$

(b) Show that

$$\Delta(x_1, \ldots, x_n) = \begin{vmatrix} n & p_1 & p_2 & \cdots & p_{n-1} \\ p_1 & p_2 & p_3 & \cdots & p_n \\ \vdots & \vdots & \vdots & & \vdots \\ p_{n-1} & p_n & p_{n+1} & \cdots & p_{2n-2} \end{vmatrix} .$$

Suggestion: multiply the underlying matrix in (a) by its transpose and take the determinant.

15.11 • Let

$$\delta (x_1, \ldots, x_n) = \prod_{i<j} (x_i - x_j) .$$

Show that for any $\alpha \in S_n$,

$$\delta \left(x_{\alpha(1)}, \ldots, x_{\alpha(n)} \right) = \operatorname{sgn} \alpha \, \delta (x_1, \ldots, x_n) .$$

15.12 • Let $f \in F[x]$ have distinct roots $\zeta_1, \ldots, \zeta_n \in F$. Prove that

$$\Delta (\zeta_1, \ldots, \zeta_n) = (-1)^{n(n-1)/2} \prod_{j=1}^{n} f' (\zeta_j) .$$

Chapter 16

Roots of Equations

Introduction

In this chapter and the following ones, we are going to discuss solving polynomial equations. For a quadratic equation, it is easy to write a formula for the roots. This formula tells you a lot about them and is very useful. There is a similar formula for the roots of a cubic. But it is more complicated, less informative, and less useful (see Chapter 20). For quartics this is even more so, and for equations of degree greater than 4, such formulas do not even exist. So in trying to describe the solutions of an equation we will take a different approach, namely the one outlined in the introduction to Chapter 14.

As we saw in Chapter 7, algebraic relations among the roots are important. First, let us be more precise about what we mean by this. To make things more concrete, suppose $f(x) = x^n + \cdots + a_1 x + a_0 \in \mathbb{Q}[x]$. According to the fundamental theorem of algebra, f has n complex roots, $\zeta_1, \ldots, \zeta_n$. Then an algebraic relation among the roots is a polynomial g in n variables with coefficients in $\mathbb{Q}$ such that $g(\zeta_1, \ldots, \zeta_n) = 0$. Examples of such relations are

$$s_j(\zeta_1, \ldots, \zeta_n) = (-1)^j a_{n-j},$$

for $1 \leq j \leq n$. These hold for any f. But there may well be others for a particular polynomial.

Now it is awkward to work directly with these relations. But there is an approach that works amazingly well. The basic idea is the following. The roots of the polynomial f lie in $\mathbb{C}$, which is a very large field. In fact they lie in a much smaller field. If we look at the smallest field containing them, we will see that its algebraic structure is determined by the algebraic relations among the roots. It is this field, called the splitting field of f, which we will study. A few examples will clarify what this means.

Examples 16.1

(i) $f(x) = x^2 - 2 \in \mathbb{Q}[x]$. The roots of f are $\pm\sqrt{2} \in \mathbb{C}$. These lie in the field $\mathbb{Q}(\sqrt{2})$. This is clearly the smallest field containing $\mathbb{Q}$ and the two roots. Elements of $\mathbb{Q}(\sqrt{2})$ are of the form $a + b\sqrt{2}$, $a, b \in \mathbb{Q}$, and the multiplication in $\mathbb{Q}(\sqrt{2})$ is determined by the relation

$$(\sqrt{2})^2 = 2 \,.$$

(ii) Let $f(x) = x^4 + x^3 + x^2 + x + 1 \in \mathbb{Q}[x]$. As we saw in Chapter 7, the roots of f are

$$\zeta_1 = e^{2\pi i/5} \,, \quad \zeta_2 = e^{4\pi i/5} \,, \quad \zeta_3 = e^{6\pi i/5} \,, \quad \zeta_4 = e^{8\pi i/5} \,.$$

These satisfy the relations determined by the symmetric polynomials. They also satisfy

$$\zeta_2 = \zeta_1^2 \,, \quad \zeta_3 = \zeta_1^3 \,, \quad \zeta_4 = \zeta_1^4$$

Let us set $\omega := \zeta_1$. Then all the roots lie in the field $\mathbb{Q}(\omega)$ (see Example 14.4(iv)), and this is the smallest field containing $\mathbb{Q}$ and the four roots. Elements of $\mathbb{Q}(\omega)$ are of the form $a_0 + a_1\omega + a_2\omega^2 + a_3\omega^3$, and multiplication in $\mathbb{Q}(\omega)$ is determined by the relation

$$-1 = s_1(\zeta_1, \zeta_2, \zeta_3, \zeta_4) = \zeta_1 + \zeta_2 + \zeta_3 + \zeta_4 = \omega + \omega^2 + \omega^3 + \omega^4 \,. \quad \square$$

In general, given $f \in \mathbb{Q}[x]$, with roots $\zeta_1, \ldots, \zeta_n \in \mathbb{C}$, let

$$\mathbb{Q}(\zeta_1, \ldots, \zeta_n) := \{g(\zeta_1, \ldots, \zeta_n) \in \mathbb{C}\} \,,$$

where $g = h/k$ is a rational function, with $h, k \in \mathbb{Q}[x_1, \ldots, x_n]$, and $k(\zeta_1, \ldots, \zeta_n) \neq 0$. It is not hard to see that $\mathbb{Q}(\zeta_1, \ldots, \zeta_n)$ is a field, and in fact,

$$\mathbb{Q}(\zeta_1, \ldots, \zeta_n) = \cap \{K \mid \mathbb{Q} \subset K \subset \mathbb{C}, \ \zeta_1, \ldots, \zeta_n \in K, \ K \text{ a field}\} \,.$$

In other words, $\mathbb{Q}(\zeta_1, \ldots, \zeta_n)$ is the smallest field containing $\zeta_1, \ldots, \zeta_n$ and $\mathbb{Q}$. The relations among the roots determine the structure of this field. However, this is too vague to be of much use. So we will develop a clearer, more precise description.

In the examples above, we have extended the field $\mathbb{Q}$ to obtain a larger field containing the roots of f. These fields are called extension fields of $\mathbb{Q}$. As was mentioned, the relative size of an extension field is important. There is a simple way of measuring this. Such a field is a $\mathbb{Q}$-vector space. So we can measure its size by its dimension. In the first example, it is 2, and in the second, 4. In general, $\mathbb{Q}(\zeta_1, \ldots, \zeta_n)$ is finite dimensional. On the other hand, the dimension of $\mathbb{C}$ as a $\mathbb{Q}$-vector space is infinite. So in this sense, $\mathbb{C}$ is a very large field, and $\mathbb{Q}(\zeta_1, \ldots, \zeta_n)$, a much smaller one.

Extension Fields

DEFINITION 16.1 *Let E and F be fields, with E $\supset$ F. Then E is called an extension field of F, and F, a subfield of E.*

One often refers to the *extension E over F*, written E/F. So $\mathbb{Q}(\sqrt{2})$ and $\mathbb{Q}(\omega)$ are extension fields of $\mathbb{Q}$ and they are subfields of $\mathbb{C}$. In Example 14.4(iii), $\mathbb{F}_p$ is embedded in $\mathbb{F}_{p^2}$ as the set of diagonal matrices:

$$\mathbb{F}_p \cong \left\{ \begin{pmatrix} a & 0 \\ 0 & a \end{pmatrix} \middle| a \in \mathbb{F}_p \right\} \subset \left\{ \begin{pmatrix} a & b \\ br & a \end{pmatrix} \middle| a, b \in \mathbb{F}_p \right\} = \mathbb{F}_{p^2} .$$

These are all examples of *simple* extensions. An extension E/F is called *simple* if there exists an element $\zeta \in E$ such that

$$E = F(\zeta) := \{ g(\zeta)/h(\zeta) \mid g, h \in F[x], h(\zeta) \neq 0 \} .$$

We say that $F(\zeta)$ is obtained from F by *adjoining* ζ. Simple extensions can be described precisely, which makes them very useful.

THEOREM 16.1
Suppose E = F(ζ), for some $\zeta \in E$. Then either

$$F(\zeta) \cong F[x]/(f) ,$$

for some irreducible polynomial f $\in$ F[x], or

$$F(\zeta) \cong F(x) ,$$

the field of rational functions in x.

PROOF   Suppose there exists $f \in F[x]$, $f \neq 0$, such that $f(\zeta) = 0$. Pick one of minimal degree. This f will be irreducible: for if $f = gh$, with $\deg g$, $\deg h < \deg f$, then either $g(\zeta) = 0$ or $h(\zeta) = 0$. We have the evaluation map (see Remark 14.4)

$$\epsilon_\zeta : F[x] \to F(\zeta)$$

given by

$$\epsilon_\zeta(g) = g(\zeta) ,$$

which induces an isomorphism

$$\bar{\epsilon}_\zeta : F[x]/(f) \to F(\zeta) .$$

The other possibility is that $f(\zeta) \neq 0$ for all $f \in F[x]$, $f \neq 0$. We define $\epsilon_\zeta : F[x] \to f(\zeta)$ as above. But now we can extend it to $F(x)$ by setting

$$\epsilon_\zeta(g/h) = g(\zeta)/h(\zeta) ,$$

for any $g, h \in F[x]$, $h \neq 0$. This gives a homomorphism, whose image again is an extension of F containing ζ. Therefore it must be all of $F(\zeta)$. ∎

In the first case, the polynomial f chosen in the proof can be taken to be monic. It is then called the *minimal polynomial* of ζ. Its degree is called the *degree* of ζ.

COROLLARY 16.1
If E/F is an extension, and $\zeta \in E$, with $g(\zeta) = 0$ for some $g \in F[x]$, then g is divisible by the minimal polynomial of ζ.

PROOF   If $g(\zeta) = 0$, then $\bar{\epsilon}_\zeta(g) = 0$, which means that $\bar{g} = 0 \in F[x]/(f)$. In other words, $g \in (f)$ or $f \mid g$. ∎

In general, if E/F is an extension and $\zeta \in E$, one says that ζ is *algebraic* over F if there exists $f \in F[x]$ such that $f(\zeta) = 0$. If no such f exists, one says that ζ is *transcendental* over F. If ζ is algebraic over $\mathbb{Q}$, then ζ is called an *algebraic number,* and if it is transcendental over $\mathbb{Q}$, a *transcendental number.* It is not hard to see that there are countably many algebraic numbers. So there are uncountably many transcendental numbers. For example, π and e are transcendental. Yet it is very hard to prove that a complex number is transcendental (see [1, Section 6.3])

Let us look at a slightly more complicated example:

Example 16.2
Let $f(x) = x^3 - 2$. Clearly $\sqrt[3]{2} \in \mathbb{R}$ is a root of f. If ω is a cube root of 1, then $\omega\sqrt[3]{2}$ is also a root of f. So let $\omega = e^{2\pi i/3}$. Then

$$\sqrt[3]{2}, \quad \omega\sqrt[3]{2}, \quad \omega^2\sqrt[3]{2} \quad \in \mathbb{C}$$

are the roots of f. We want to describe $\mathbb{Q}(\sqrt[3]{2}, \omega\sqrt[3]{2}, \omega^2\sqrt[3]{2})$. Now

$$\mathbb{Q}\left(\sqrt[3]{2}\right) \subset \mathbb{R} ,$$

but

$$\omega\sqrt[3]{2}, \ \omega^2\sqrt[3]{2} \notin \mathbb{R}$$

So

$$\omega\sqrt[3]{2}, \ \omega^2\sqrt[3]{2} \notin \mathbb{Q}\left(\sqrt[3]{2}\right) .$$

Any field containing all three roots must contain

$$\omega\sqrt[3]{2}/\sqrt[3]{2} = \omega .$$

So let us adjoin ω to $\mathbb{Q}(\sqrt[3]{2})$. Then

$$E := \mathbb{Q}\left(\sqrt[3]{2}\right)(\omega) = \mathbb{Q}\left(\sqrt[3]{2}, \omega\right) \subset \mathbb{Q}\left(\sqrt[3]{2}, \omega\sqrt[3]{2}, \omega^2\sqrt[3]{2}\right),$$

and on the other hand, E will contain $\omega\sqrt[3]{2}$ and $\omega^2\sqrt[3]{2}$. So

$$\mathbb{Q}\left(\sqrt[3]{2}, \omega\sqrt[3]{2}, \omega^2\sqrt[3]{2}\right) = \mathbb{Q}\left(\sqrt[3]{2}, \omega\right).$$

Can you describe $\mathbb{Q}(\sqrt[3]{2}, \omega)$ as a simple extension? $\square$

In general, suppose that E/F is a field extension, and $\zeta_1, \ldots, \zeta_r \in E$. Then we can define the extension

$$\begin{aligned}
F(\zeta_1, \ldots, \zeta_r) &= F(\zeta_1) \ldots (\zeta_r) \subset E \\
&= \cap \{K \mid F \subset K \subset E, \ \zeta_1, \ldots, \zeta_r \in K\} \\
&= \{g(\zeta_1, \ldots, \zeta_r) \in E\},
\end{aligned}$$

where $g = h/k$ is a rational function, with $h, k \in F[x_1, \ldots, x_n]$ and $k(\zeta_1, \ldots, \zeta_n) \neq 0$. We say that $F(\zeta_1, \ldots, \zeta_r)$ is obtained from F by adjoining $\zeta_1, \ldots, \zeta_r$.

Degree of an Extension

If E is an extension of F, then E is an F-vector space:

(i) for $\zeta, \eta \in E$, define vector addition to be field addition in E

$$\zeta + \eta := \zeta + \eta \in E.$$

(ii) for $a \in F$ and $\zeta \in E$, define scalar multiplication using the field multiplication in E

$$a\zeta := a\zeta \in E.$$

It is straightforward to check that E, with these two operations, is an F-vector space.

DEFINITION 16.2   *Let E/F be a field extension. Then the degree of the extension, written $[E : F]$, is the dimension of E as an F-vector space, that is,*

$$[E : F] := \dim_F E.$$

If $[E : F] < \infty$ then E/F is called a finite extension.

If $f \in F[x]$ is irreducible, then we saw in the previous chapter that the dimension of the simple extension E constructed from f is $\deg f$. So in this case,

$$[E : F] = \deg f .$$

In our first example,

$$[\mathbb{Q}(\sqrt{2}) : \mathbb{Q}] = 2 ,$$

with basis

$$\{1, \sqrt{2}\} .$$

In the second one,

$$[\mathbb{Q}(\omega) : \mathbb{Q}] = 4 ,$$

where $\omega = e^{2\pi i/5}$, and

$$\left\{1, \omega, \omega^2, \omega^3\right\}$$

is a basis. In the third example,

$$\left[\mathbb{F}_{p^2} : \mathbb{F}_p\right] = 2$$

and

$$\left\{1, \begin{pmatrix} 0 & 1 \\ r & 0 \end{pmatrix}\right\}$$

is a basis.

In Example 16.2, we have

$$\mathbb{Q} \subset \mathbb{Q}\left(\sqrt[3]{2}\right) \subset \mathbb{Q}\left(\sqrt[3]{2}, \omega\right) .$$

It is clear that

$$[\mathbb{Q}(\sqrt[3]{2}) : \mathbb{Q}] = 3$$

with basis

$$\left\{1, \sqrt[3]{2}, \left(\sqrt[3]{2}\right)^2\right\} .$$

What about $[\mathbb{Q}(\sqrt[3]{2}, \omega) : \mathbb{Q}(\sqrt[3]{2})]$? Well, ω is a root of x^2+x+1, which is irreducible in $\mathbb{R}[x]$. Therefore it is irreducible over $\mathbb{Q}(\sqrt[3]{2})$ and

$$[\mathbb{Q}(\sqrt[3]{2}, \omega) : \mathbb{Q}(\sqrt[3]{2})] = 2$$

with basis

$$\{1, \omega\} .$$

We would also like to compute $[\mathbb{Q}(\sqrt[3]{2}, \omega) : \mathbb{Q}]$. It is not obvious that $\mathbb{Q}(\sqrt[3]{2}, \omega)/\mathbb{Q}$ is a simple extension. But the following result shows that we can calculate its degree from $[\mathbb{Q}(\sqrt[3]{2}, \omega) : \mathbb{Q}(\sqrt[3]{2})]$ and $[\mathbb{Q}(\sqrt[3]{2}) : \mathbb{Q}]$.

THEOREM 16.2
Let $D \supset E \supset F$ be fields. Then

$$[D : F] = [D : E][E : F] .$$

PROOF First, let us assume that $[D : E]$ and $[E : F]$ are both finite. So suppose that

$$S = \{\zeta_1, \dots , \zeta_s\}$$

is a basis of D over E, and

$$T = \{\eta_1, \dots , \eta_t\} ,$$

a basis of E over F. Set

$$U = \left\{\zeta_j \eta_k \in D \mid 1 \leq j \leq s, 1 \leq k \leq t\right\} .$$

We want to show that U is a basis of D/F. First we show that it generates D/F. Pick $\theta \in D$. Since S is a set of generators of D over E, there exist $\alpha_1, \dots \alpha_s \in E$ such that

$$\theta = \alpha_1 \zeta_1 + \cdots + \alpha_s \zeta_s .$$

Now T generates E over F. So for each j, $1 \leq j \leq s$, there exist $\beta_{jk} \in F, 1 \leq k \leq t$, such that

$$\alpha_j = \beta_{j1} \eta_1 + \cdots + \beta_{jt} \eta_t .$$

Therefore

$$\theta = \sum_{j=1}^{s} \sum_{k=1}^{t} \beta_{jk} \left(\zeta_j \eta_k\right) .$$

So U generates D over F.

Now we check that U is linearly independent over F. Suppose that

$$0 = \sum_{j=1}^{s} \sum_{k=1}^{t} \beta_{jk} \left(\zeta_j \eta_k\right) = \sum_{j=1}^{s} \left(\sum_{k=1}^{t} \beta_{jk} \eta_k\right) \zeta_j$$

for some $\beta_{jk} \in F, 1 \leq j \leq s, 1 \leq k \leq t$. Since S is linearly independent over E, the coefficients of $\zeta_1, \dots , \zeta_s$ must be 0, in other words,

$$\beta_{j1} \eta_1 + \cdots + \beta_{jt} \eta_t = 0 ,$$

for $1 \leq j \leq s$. But $\{\eta_1, \dots , \eta_t\}$ are linearly independent over F. Therefore $\beta_{jk} = 0$ for all j, k. So U is linearly independent over F and is therefore a basis of D/F. It follows that

$$[D : E][E : F] = [D : F] ,$$

if all three are finite.

If one of $[D : E]$ or $[E : F]$ is infinite, then $[D : F]$ is infinite. ∎

We now can compute $[\mathbb{Q}(\sqrt[3]{2}, \omega) : \mathbb{Q}]$:

$$[\mathbb{Q}(\sqrt[3]{2}, \omega) : \mathbb{Q}] = [\mathbb{Q}(\sqrt[3]{2}, \omega) : \mathbb{Q}(\sqrt[3]{2})][\mathbb{Q}(\sqrt[3]{2}) : \mathbb{Q}] = 2 \cdot 3 = 6 .$$

Using the bases for $\mathbb{Q}(\sqrt[3]{2})/\mathbb{Q}$ and $\mathbb{Q}(\sqrt[3]{2}, \omega)/\mathbb{Q}(\sqrt[3]{2})$ given above, the bases constructed in the theorem is

$$\left\{1, \ \sqrt[3]{2}, \ (\sqrt[3]{2})^2, \ \omega, \ \omega\sqrt[3]{2}, \ \omega(\sqrt[3]{2})^2\right\} .$$

Example 16.3
Take $f(x) = x^4 - 10x^2 + 1$ (see Chapter 7). Its roots are

$$\sqrt{2} + \sqrt{3}, \quad -\sqrt{2} + \sqrt{3}, \quad \sqrt{2} - \sqrt{3}, \quad -\sqrt{2} - \sqrt{3} .$$

Any subfield of $\mathbb{R}$ containing the four roots contains $\sqrt{2}$ and $\sqrt{3}$ and vice versa. So the smallest field containing the roots is $\mathbb{Q}(\sqrt{2}, \sqrt{3})$. From the theorem,

$$[\mathbb{Q}(\sqrt{2}, \sqrt{3}) : \mathbb{Q}] = [\mathbb{Q}(\sqrt{2}) : \mathbb{Q}][\mathbb{Q}(\sqrt{2}, \sqrt{3}) : \mathbb{Q}(\sqrt{2})] = 2 \cdot 2 = 4 ,$$

since $x^2 - 3$ has no root in $\mathbb{Q}(\sqrt{2})$.

In fact it is not hard to see that $\mathbb{Q}(\sqrt{2}, \sqrt{3})/\mathbb{Q}$ is a simple extension. We have

$$\mathbb{Q}(\sqrt{2} + \sqrt{3}) \subset \mathbb{Q}(\sqrt{2}, \sqrt{3}) .$$

A short calculation shows that we can express $\sqrt{2}$ and $\sqrt{3}$ in terms of $\sqrt{2} + \sqrt{3}$ and its powers:

$$(\sqrt{2} + \sqrt{3})^2 = 5 + 2\sqrt{2}\sqrt{3} ,$$

and

$$(\sqrt{2} + \sqrt{3})^3 = 11\sqrt{2} + 9\sqrt{3} .$$

Thus

$$\sqrt{2} = -\frac{9}{2}(\sqrt{2} + \sqrt{3}) + \frac{1}{2}(\sqrt{2} + \sqrt{3})^3 ,$$

and

$$\sqrt{3} = \frac{11}{2}(\sqrt{2} + \sqrt{3}) - \frac{1}{2}(\sqrt{2} + \sqrt{3})^3 .$$

Therefore

$$\mathbb{Q}(\sqrt{2} + \sqrt{3}) = \mathbb{Q}(\sqrt{2}, \sqrt{3}) . \quad \Box$$

Splitting Fields

We can now say more about the smallest extension field containing the roots of a polynomial. Suppose that F is a field, and f a polynomial of degree n with coefficients in F. We say that f *splits* in an extension E of F if f has n roots $\zeta_1, \ldots, \zeta_n \in E$, or equivalently,

$$f(x) = (x - \zeta_1) \cdots (x - \zeta_n) \in E[x] .$$

We want to show that any polynomial $f \in F[x]$ splits in some extension of F. The smallest such extension is called a *splitting field* of f. This is the smallest field containing all the roots of f.

For example,

(i) $\mathbb{Q}(\sqrt{2})$ is the splitting field of $x^2 - 2 \in \mathbb{Q}[x]$;

(ii) $\mathbb{Q}(\omega)$, where $\omega = e^{2\pi i/5}$, is the splitting field of $x^4 + x^3 + x^2 + x + 1 \in \mathbb{Q}[x]$;

(iii) $\mathbb{Q}(\sqrt[3]{2}, \omega)$ is the splitting field of $x^3 - 2 \in \mathbb{Q}[x]$;

(iv) $\mathbb{Q}(\sqrt{2}, \sqrt{3})$ is the splitting field of $x^4 - 10x^2 + 1 \in \mathbb{Q}[x]$.

As mentioned at the beginning of the chapter, the structure of the splitting field precisely reflects the algebraic relations among the roots of the polynomial. In the next chapter, we will use splitting fields to study the symmetries of the roots.

An obvious question to ask is, does a splitting field for a polynomial $f \in F[x]$ always exist? If the answer is yes, then we can ask whether it is unique. In examples (i) and (ii) above, we were able to construct a splitting field by adjoining one root to the field F. In example (iii), we had to adjoin a second root. This suggests how we can construct splitting fields in general. Uniqueness of splitting fields will be discussed in the next chapter.

THEOREM 16.3
Let $f \in F[x]$ be a polynomial of degree n. There exists a splitting field E/F of f with $[E : F] \leq n!$.

PROOF We first assume that f is irreducible and argue by induction on n. If $n = 1$, then f is linear and its only root lies in F. Suppose that the result holds for irreducible polynomials of degree at most $n - 1$. Set

$$E = F[x]/(f) .$$

In E, f has at least one root. So we can write

$$f(x) = (x - \zeta_1) \cdots (x - \zeta_r) g(x) \in E[x] ,$$

with $\zeta_1, \ldots, \zeta_r \in E$, and $g \in E[x]$ irreducible. Since $\deg g < n$, by the induction assumption, the result applies to g. Thus there exists a splitting field D/E for g with

$$[D : E] \leq (n - r)! .$$

If

$$g(x) = (x - \zeta_{r+1}) \cdots (x - \zeta_n) \in D[x] ,$$

then

$$f(x) = (x - \zeta_1) \cdots (x - \zeta_n) \in D[x],$$

in other words, f splits in D. The field D then contains a splitting field E' of f with

$$[E' : F] \le [D : F] = [D : E][E : F] \le (n - r)!n \le n!,$$

as desired. If f is reducible, we can apply the result for irreducible polynomials to each irreducible factor of f. It is not hard to see that here too, the degree of the splitting field is at most $n!$. ∎

Example 16.4

Consider $f(x) = x^p - 2$, where p is prime. We have already looked at $p = 3$. First, there is a real root, $\sqrt[p]{2}$. Furthermore, if $\omega^p = 1$, then $\omega \sqrt[p]{2}$ is also a root of f. So pick a pth root of 1, $\omega \neq 1$, for example $\omega = e^{2\pi i/p}$. Then the remaining pth roots of 2 are $\omega^j \sqrt[p]{2}$, $1 < j < p$. As in the case $p = 3$, we see that if an extension field contains all the roots, then it must contain

$$\omega \sqrt[p]{2} / \sqrt[p]{2} = \omega.$$

Conversely, if it contains $\sqrt[p]{2}$ and ω, then it contains all the roots. So the splitting field of f is

$$\mathbb{Q}(\sqrt[p]{2}, \omega).$$

To calculate $[\mathbb{Q}(\sqrt[p]{2}, \omega) : \mathbb{Q}]$, first note that

$$[\mathbb{Q}(\sqrt[p]{2}) : \mathbb{Q}] = p.$$

So

$$p \mid [\mathbb{Q}(\sqrt[p]{2}) : \mathbb{Q}][\mathbb{Q}(\sqrt[p]{2}, \omega) : \mathbb{Q}(\sqrt[p]{2})] = [\mathbb{Q}(\sqrt[p]{2}, \omega) : \mathbb{Q}].$$

And since ω is a root of $x^{p-1} + \cdots + x + 1$,

$$[\mathbb{Q}(\sqrt[p]{2}, \omega) : \mathbb{Q}(\sqrt[p]{2})] \le p - 1.$$

Therefore

$$[\mathbb{Q}(\sqrt[p]{2}, \omega) : \mathbb{Q}] \le (p - 1)p.$$

On the other hand,

$$[\mathbb{Q}(\omega) : \mathbb{Q}] = p - 1$$

since $x^{p-1} + \cdots + x + 1$ is the minimal polynomial of ω over $\mathbb{Q}$, and

$$[\mathbb{Q}(\omega) : \mathbb{Q}] \mid [\mathbb{Q}(\sqrt[p]{2}, \omega) : \mathbb{Q}].$$

Therefore both p and $p - 1$ divide $[\mathbb{Q}(\sqrt[p]{2}, \omega) : \mathbb{Q}]$. Since they are relatively prime, we have

$$(p - 1)p \mid [\mathbb{Q}(\sqrt[p]{2}, \omega) : \mathbb{Q}].$$

Thus
$$[\mathbb{Q}(\sqrt[p]{2}, \omega) : \mathbb{Q}] = (p - 1)p$$

The diagram below shows the relationships between the four fields, and the degrees of the extensions.

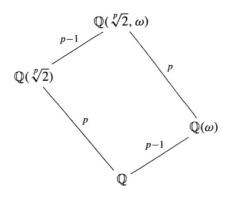

☐

Cubics

Let F be a field of characteristic $\neq 2$ and $f(x) = x^2 + a_1 x + a_0 \in F[x]$, a quadratic polynomial. On page 215, we recalled the standard formula for its roots ζ_1 and ζ_2. You can also write this in a slightly different way:

$$\zeta_1 = \frac{-a_1 + \delta}{2}, \qquad \zeta_2 = \frac{-a_1 - \delta}{2},$$

where $\delta = \sqrt{a_1^2 - 4a_0}$ is a square root of the discriminant of f. This formula tells us that
$$\delta = \zeta_1 - \zeta_2,$$

and that $F(\delta)$ is a splitting field of f over F.

Now suppose that
$$f = x^3 + a_2 x^2 + a_1 x + a_0 \in F[x]$$

is a cubic with 3 distinct roots ζ_1, ζ_2, and ζ_3 in some extension E/F. So we can write
$$f(x) = (x - \zeta_1)(x - \zeta_2)(x - \zeta_3) \in E[x].$$

Set
$$\delta := (\zeta_1 - \zeta_2)(\zeta_2 - \zeta_3)(\zeta_1 - \zeta_3) \in E$$

So $\Delta = \delta^2$ is the discriminant of f. It is not hard to see that $\Delta \in F$. If we pick one root, say ζ_1, then we can express the other two in terms of it and δ. First, we have

$$-a_2 = \zeta_1 + \zeta_2 + \zeta_3 \,,$$

which gives us

$$\zeta_2 + \zeta_3 = -a_2 - \zeta_1 \,. \qquad (16.1)$$

Furthermore,

$$f'(x) = (x - \zeta_2)(x - \zeta_3) + (x - \zeta_1)(x - \zeta_3) + (x - \zeta_1)(x - \zeta_2) \,,$$

so that

$$f'(\zeta_1) = (\zeta_1 - \zeta_2)(\zeta_1 - \zeta_3) \,.$$

Therefore

$$\zeta_2 - \zeta_3 = \frac{\delta}{f'(\zeta_1)} \,.$$

Combining this with Equation (16.1), we get

$$\begin{aligned} \zeta_2 &= -\frac{a_2}{2} - \frac{\zeta_1}{2} + \frac{\delta}{2f'(\zeta_1)} \\ \zeta_3 &= -\frac{a_2}{2} - \frac{\zeta_1}{2} - \frac{\delta}{2f'(\zeta_1)} \end{aligned} \qquad (16.2)$$

Thus $F(\delta, \zeta_1)$ is a splitting field for f over F. One should think of this as $F(\delta)$ with a root of f adjoined. If f is irreducible over F, then

$$\big[F(\delta, \zeta_1) : F\big] = 3 \text{ or } 6 \,,$$

depending upon whether $\delta \in F$ or $\delta \notin F$.

For example, take $f(x) = x^3 - 2 \in \mathbb{Q}[x]$. As we saw in Example 16.2, the roots are

$$\zeta_1 = \sqrt[3]{2} \,, \quad \zeta_2 = \omega\sqrt[3]{2} \,, \quad \zeta_3 = \omega^2\sqrt[3]{2} \,,$$

where $\omega = e^{2\pi i/3}$. So

$$\begin{aligned} \delta &= (\sqrt[3]{2} - \omega\sqrt[3]{2})(\omega\sqrt[3]{2} - \omega^2\sqrt[3]{2})(\sqrt[3]{2} - \omega^2\sqrt[3]{2}) \\ &= 2(1 - \omega)(\omega - \omega^2)(1 - \omega^2) \\ &= 6 + 12\omega \,, \end{aligned}$$

and $\mathbb{Q}(\delta, \zeta_1) = \mathbb{Q}(\omega, \sqrt[3]{2})$ is the splitting field again.

REMARK 16.1   If f is a real quadratic then the discriminant controls whether the roots are real: if Δ is positive, then f has two real roots. If Δ is negative, then f has a pair of complex conjugate roots. For a cubic $f \in \mathbb{R}[x]$, the discriminant has a similar significance. A real cubic always has a real root. So take ζ_1 to be this root. Then Equation (16.2) shows that the other two roots are real if the discriminant is positive. They are complex conjugate if the discriminant is negative. The discriminant of real cubics is discussed further in the section "Plots and Calculations" in this chapter.

∎

Cyclotomic Polynomials

We have been making much use of roots of unity. In this section, we are going to discuss their minimal polynomials, called cyclotomic polynomials.

As we saw in Exercise 6.2, the nth roots of 1 form a cyclic group μ_n of order n. A *primitive* nth root of 1 is one that generates the group. If we write μ_n as

$$\left\{ e^{2\pi i k/n} \mid 0 < k \leq n \right\} \subset \mathbb{C},$$

then the primitive roots are those with $(k, n) = 1$. Thus there are $\varphi(n)$ of them.

Here is a plot of the 18th roots of 1 and the primitive 18th roots of 1 in the complex plane.

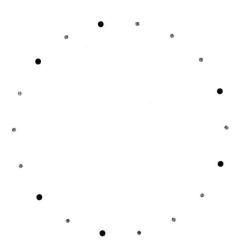

We define the nth cyclotomic polynomial $\Phi_n(x)$ to be the monic polynomial whose roots are the primitive nth roots of 1, that is

$$\Phi_n(x) = \prod_{\substack{\omega \text{ primitive} \\ n\text{th root of } 1}} (x - \omega) = \prod_{\substack{(k,n)=1 \\ 0<k<n}} \left(x - e^{2\pi i k/n} \right).$$

If ω is an nth root of 1, and the order of ω in μ_n is d, then $d \mid n$ and ω is a primitive dth root of 1. Conversely, if $d \mid n$ and ω is a primitive dth root of 1, then ω is an nth root of 1 whose order is d. Therefore, sorting the roots by their order, we see that

$$x^n - 1 = \prod_{\omega^n=1} (x - \omega) = \prod_{d \mid n} \prod_{\substack{\omega \text{ primitive} \\ d\text{th root of } 1}} (x - \omega) = \prod_{d \mid n} \Phi_d(x). \tag{16.3}$$

Using this formula, we can compute the cyclotomic polynomials recursively. We have

$$x - 1 = \Phi_1(x),$$

and

$$x^2 - 1 = \Phi_1(x)\Phi_2(x) = (x - 1)\Phi_2(x).$$

Therefore

$$\Phi_2(x) = x + 1.$$

Next

$$x^3 - 1 = \Phi_1(x)\Phi_3(x) = (x - 1)\Phi_3(x).$$

So

$$\Phi_3(x) = x^2 + x + 1.$$

In fact, for any prime p,

$$x^p - 1 = (x - 1)\Phi_p(x),$$

and thus

$$\Phi_p(x) = x^{p-1} + \cdots + x + 1.$$

Going on, we have

$$x^4 - 1 = \Phi_1(x)\Phi_2(x)\Phi_4(x) = (x - 1)(x + 1)\Phi_4(x).$$

Therefore

$$\Phi_4(x) = x^2 + 1.$$

And

$$x^6 - 1 = \Phi_1(x)\Phi_2(x)\Phi_3(x)\Phi_6(x) = (x - 1)(x + 1)(x^2 + x + 1)\Phi_6(x),$$

which tells us that

$$\Phi_6(x) = x^2 - x + 1.$$

You can continue like this. You can also use software to compute Φ_n for much larger values of n. Looking at these calculations, the first thing you notice is that all the cyclotomic polynomials computed have integer coefficients.

THEOREM 16.4
The cyclotomic polynomials Φ_n have integer coefficients and are monic, for all $n \in \mathbb{N}$.

PROOF   To see this, we argue by induction on n. It is certainly true for $n = 1$. Suppose that all Φ_d for $d < n$, have integer coefficients and are monic. Therefore

$$g(x) := \prod_{\substack{d|n \\ d<n}} \Phi_d(x)$$

is monic with integer coefficients. By Equation (16.3), we have

$$x^n - 1 = g(x)\Phi_n(x) \ .$$

It follows that Φ_n too must have integer coefficients and be monic (see Exercise 14.8). So by the principle of induction, the first statement is proved. ∎

REMARK 16.2 Calculations of Φ_n for small n may lead you to ask whether its coefficients are ± 1 for all n. This is not the case. For example,

$$\Phi_{105}(x) = 1 + x + x^2 - x^5 - x^6 - 2x^7 - x^8 - x^9 + x^{12} + x^{13} + x^{14} + x^{15} + x^{16}$$
$$+ x^{17} - x^{20} - x^{22} - x^{24} - x^{26} - x^{28} + x^{31} + x^{32} + x^{33} + x^{34} + x^{35}$$
$$+ x^{36} - x^{39} - x^{40} - 2x^{41} - x^{42} - x^{43} + x^{46} + x^{47} + x^{48} \ .$$

In fact arbitrarily large and small integers occur as coefficients of cyclotomic polynomials. ∎

Looking again at our calculations of cyclotomic polynomials, you notice that Φ_3, Φ_4 and Φ_6 are irreducible over $\mathbb{Q}$. In Example 14.2, we saw that Φ_p is irreducible. So one might ask whether this is true for all Φ_n.

THEOREM 16.5
The cyclotomic polynomials Φ_n are irreducible over $\mathbb{Q}$, for all $n \in \mathbb{N}$.

PROOF Suppose that $f \in \mathbb{Q}[x]$ is monic, irreducible, and divides Φ_n. It follows from Exercise 14.8 that f has integer coefficients. Since Φ_n divides $x^n - 1$, so does f. Write

$$x^n - 1 = f(x)g(x) \ ,$$

for some g with integer coefficients. Now let ω be a primitive nth root of 1, which is a root of f. Thus f is the minimal polynomial of ω. Pick a prime p that does not divide n. Then
$$0 = (\omega^p)^n - 1 = f(\omega^p)g(\omega^p) \ .$$
So either $f(\omega^p) = 0$ or $g(\omega^p) = 0$.

Suppose that $g(\omega^p) = 0$. Then ω is a root of $g(x^p)$. It follows from Corollary 16.1 that $f(x) \mid g(x^p)$, say

$$g(x^p) = f(x)h(x) \ ,$$

where h also has integer coefficients. Next, reduce this equation mod p:

$$\bar{g}(x^p) = \bar{f}(x)\bar{h}(x) \ .$$

Since $\bar{g}(x^p) = \bar{g}(x)^p$ (see Exercise 14.21), we have

$$\bar{g}(x)^p = \bar{f}(x)\bar{h}(x) \ .$$

Now in $\mathbb{F}_p[x]$, every polynomial factors uniquely into a product of irreducibles (Theorem 14.3). Therefore, some irreducible factor $k \in \mathbb{F}_p[x]$ of $\bar{g}$ is also a factor of $\bar{f}$. Hence,

$$k^2 \mid \bar{f}\bar{g} = \bar{x}^n - 1 \, .$$

But if $\bar{x}^n - 1$ has a repeated factor, then by Exercise 14.20,

$$(\bar{x}^n - 1, n\bar{x}^{n-1}) \neq 1 \, ,$$

which is impossible. It follows that ω^p must be a root of f for every prime $p \nmid n$.

We want to see that ω^r is a root of f for every r relatively prime to n. Suppose that $r = p_1 \cdots p_l$, where $p_1, \dots, p_l$ are primes that do not divide n. Now in the argument above, we can replace ω by ω^{p_1}, since it too is a primitive nth root of 1. It follows that $(\omega_1^p)^{p_2} = \omega^{p_1 p_2}$ is a root of f. Continuing like this we see that ω^r is as well. Since every root of Φ_n is of this form, we must have that $f = \Phi_n$, and so Φ_n is irreducible. ∎

Since μ_n is a cyclic group, an extension of $\mathbb{Q}$ that contains one primitive nth root contains all of them. So the splitting field E_n of Φ_n is isomorphic to $\mathbb{Q}[x]/(\Phi_n)$. It contains all nth roots of 1 and is thus the splitting field of $x^n - 1$ as well. Furthermore,

$$\left[E_n : \mathbb{Q}\right] = \varphi(n) \, .$$

E_n is called a *cyclotomic* field.

Finite Fields

Let us now look more closely at the case where F is a field of characteristic p, in particular where F is finite with q elements. Suppose that E/F is an extension of degree r. For example, if $f \in F[x]$ is irreducible of degree r, then $F[x]/(f)$ is such a field. Since $[E : F] = r$,

$$|E| = q^r$$

(see Exercise 14.15). It follows that

$$|E^\times| = q^r - 1 \, .$$

So if $\zeta \in E^\times$,

$$\zeta^{q^r - 1} = 1 \, .$$

Therefore, for all $\zeta \in E$, (cf. Fermat's Little Theorem)

$$\zeta^{q^r} = \zeta .$$

This tells us that $x^{q^r} - x$ has q^r roots in E. So E is in fact a splitting field for $x^{q^r} - x$ over F. We are going to show in the next chapter that splitting fields are unique up to isomorphism. This means that all extensions E/F of degree r are isomorphic. So given the uniqueness result, we have:

THEOREM 16.6
There is a unique field of order p^r, for p prime, $r \in \mathbb{N}$.

For example, $\mathbb{F}_{p^2}$ is the field with p^2 elements and is the splitting field over $\mathbb{F}_p$ of $x^{p^2} - x$. In general, let us denote by $\mathbb{F}_{p^r}$, *the* field with p^r elements. If $r \mid s$, then there is an extension of degree s/r of $\mathbb{F}_{p^r}$. It has $p^{r(s/r)} = p^s$ elements, and is therefore just $\mathbb{F}_{p^s}$. The converse also holds: if $\mathbb{F}_{p^s}$ is an extension of $\mathbb{F}_{p^r}$, then $r \mid s$.

This discussion also tells us something about the irreducible polynomials over $\mathbb{F}_p$. Suppose $f \in \mathbb{F}_p$ is monic and irreducible of degree r, and $r \mid s$. Since $\mathbb{F}_p[x]/(f) \cong \mathbb{F}_{p^r}$, f is the minimal polynomial of some element $\zeta \in \mathbb{F}_{p^r}$. As $\mathbb{F}_{p^r} \subset \mathbb{F}_{p^s}$,

$$\zeta^{p^s} - \zeta = 0 .$$

Therefore by Corollary 16.1,

$$f \mid (x^{p^s} - x) .$$

Since any two distinct monic, irreducible polynomials are relatively prime, it follows that

$$\prod_{\substack{f \text{ irreducible} \\ f \text{ monic} \\ \deg f \mid s}} f \ \bigg| \ (x^{p^s} - x) .$$

Conversely, suppose that f is a monic, irreducible factor of degree r of $x^{p^s} - x$. Since $x^{p^s} - x$ splits in $\mathbb{F}_{p^s}$, so does f. Pick a root of ζ of f. Then

$$\mathbb{F}_{p^r} \cong \mathbb{F}_p[x]/(f) \cong \mathbb{F}_p(\zeta) \subset \mathbb{F}_{p^s} .$$

So $r \mid s$ and all the factors of $x^{p^s} - x$ occur in the product above. Since the leading coefficient of the product is 1, we have the result:

THEOREM 16.7

$$x^{p^s} - x = \prod_{\substack{f \text{ irreducible} \\ f \text{ monic} \\ \deg f \mid s}} f .$$

COROLLARY 16.2

Let $N(p, r)$ be the number of irreducible monic polynomials in $\mathbb{F}_p[x]$ of degree r. Then

$$p^s = \sum_{r \mid s} N(p, r)r .$$

PROOF   The degree of the left-hand side of the equation in the theorem is p^s. The degree of the right-hand side is $\sum_{r \mid s} N(p, r)r$.   ∎

You can obtain an explicit formula for $N(p, r)$ by using the Möbius inversion formula (see [1, p. 113]).

Plots and Calculations

As we have seen, the discriminant Δ of a real cubic

$$f(x) = x^3 + ax + b$$

is given by

$$\Delta = -4a^3 - 27b^2 .$$

Here is a plot of the semicubical parabola $\Delta = 0$. The regions $\Delta > 0$, respectively $\Delta < 0$, correspond to values of a and b for which f has three real roots, respectively one real root and a pair of complex conjugate roots. The package `ImplicitPlot` is used.

```
In[1]:= « Graphics`ImplicitPlot`
```

```
In[2]:= ImplicitPlot[ -4a^3 - 27b^2 == 0, {a,-2,0}]
```

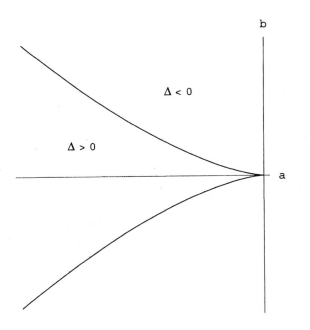

Out [2] = - Graphics -

The notebook "CubicRoots.nb" contains a demonstration showing how the discriminant controls the behavior of the roots of a 1-parameter family of real cubics. First choose a real cubic with two real critical points, for example $x^3 - 3x$. Plot its graph:

In [3] := Plot[x^3 - 3x, {x,-2,2}]

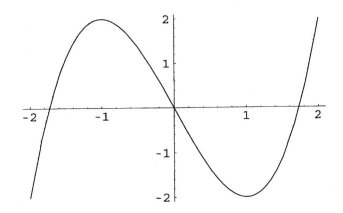

```
Out [3] = - Graphics -
```

The critical values are 2 and −2. Any value greater than 2 is taken on at one real point and two complex conjugate points. A value between 2 and −2 is taken on at three real points and one less than −2 at one real and two complex conjugate points. Equivalently one can look at the roots of $x^3 - 3x + a = 0$. In the animation below they are plotted, in black, in the complex plane for a running from -3 to 3. The discriminant $-a^2 + 4$ is also shown, in grey. One sees that when it is negative, for a between -2 and 2, there is a pair of complex conjugate roots and one real root. When it is positive, for a greater than 2 or less than −2 there are three real roots. And when it vanishes there is a double root.

```
In[4] := Do[ CubicRoots[ f, x,
            PlotRange->{{-6,6},{-1,1}} ] ], {a,-3,3,0.2} ]
```

```
Out [4] = - Graphics -
```

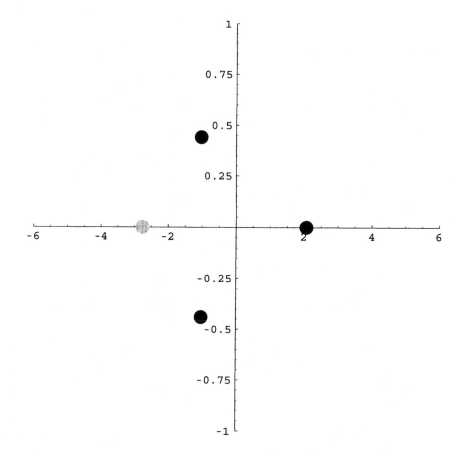

This animation looks best when run cyclically.

Mathematica has a built-in function `EulerPhi[n]` that computes $\varphi(n)$:

```
In[5]:= EulerPhi[105]
```

```
Out[5]= 48
```

The function `Cyclotomic[n, x]` calculates the nth cyclotomic polynomial $\Phi_n(x)$. Here is the example given in Remark 16.2:

```
In[6]:= Cyclotomic[105,x]
```

$$Out[6] = 1 + x + x^2 - x^5 - x^6 - 2x^7 - x^8 - x^9 + x^{12} + x^{13} + x^{14} + x^{15} + x^{16} + x^{17} - x^{20} - x^{22} - x^{24} - x^{26} - x^{28} + x^{31} + x^{32} + x^{33} + x^{34} + x^{35} + x^{36} - x^{39} - x^{40} - 2x^{41} - x^{42} - x^{43} + x^{46} + x^{47} + x^{48}$$

Exercises

16.1 • Show that if $f \in \mathbb{C}[x]$ has degree 2, then f splits in $\mathbb{C}$. Conclude that $\mathbb{C}$ has no quadratic extensions and that every quadratic extension of $\mathbb{R}$ is isomorphic to $\mathbb{C}$.

16.2 Find the splitting field of $x^3 + 2 \in \mathbb{Q}[x]$.

16.3 Find the minimal polynomial of $\sqrt{2 + \sqrt{2}}$ over $\mathbb{Q}$.

16.4 • What is the minimal polynomial of $\cos(2\pi/5)$ over $\mathbb{Q}$?

16.5 Describe $\mathbb{Q}(\sqrt[3]{2}, \omega)/\mathbb{Q}$, where $\omega \neq 1$ is a cube root of 1, as a simple extension.

16.6 Show that the cyclotomic field $E_8 = \mathbb{Q}(i, \sqrt{2})$.

16.7 • Find the roots of $x^4 - 2x^2 - 4 \in \mathbb{Q}[x]$. Determine its splitting field $E/\mathbb{Q}$. What is $[E : \mathbb{Q}]$?

16.8 • Suppose that E/F is a field extension, and let

$$K = \{\zeta \in E \mid \zeta \text{ algebraic over } F\}.$$

Prove that K is a field.

16.9 • Let $E/\mathbb{R}$ be a finite extension.

(a) Suppose that $\zeta \in E$. Show that $[\mathbb{R}(\zeta) : \mathbb{R}]$ is even or 1.

(b) Prove that $[E : \mathbb{R}]$ is even or 1.

16.10 • Let
$$f(x) = x^n + a_{n-1} x^{n-1} + \cdots + a_0 \in F[x],$$

where the characteristic of F does not divide n and $n \geq 2$. Show that the substitution
$$x = y - a_{n-1}/n$$

transforms f into a polynomial of the form
$$g(y) = x^n + b_{n-2} x^{n-2} + \cdots + b_0 \in F[x].$$

Compute the coefficients of the transformed polynomial g in the cases $n = 2$ and $n = 3$. This substitution is called a (linear) *Tschirnhausen transformation*.

16.11 • Show that the discriminant of
$$g(y) = y^3 + b_1 y + b_0$$

is
$$-4b_1^3 - 27b_0^2.$$

16.12 • Let $f \in \mathbb{R}[x]$ be a cubic polynomial with distinct roots. Prove that

(a) f has three real roots if and only if $\Delta > 0$;

(b) f has one real root and a pair of complex conjugate roots if and only if $\Delta < 0$.

16.13 Compute $\Phi_n(x)$ for $n \leq 12$.

16.14 If $p > 2$ is prime, show that
$$\Phi_{2p}(x) = x^{p-1} - x^{p-2} + \cdots + x^2 - x + 1.$$

16.15 Prove that the discriminant of Φ_p is $(-1)^{p(p-1)/2} p^{p-2}$. Suggestion: write
$$x^p - 1 = (x - 1)\Phi_p(x),$$

and differentiate to obtain
$$px^{p-1} = \Phi_p(x) + (x - 1)\Phi'_p(x).$$

Then apply the result of Exercise 15.12.

16.16 Prove that

$$\sum_{d|n} \varphi(d) = n .$$

16.17 Show that if $\mathbb{F}_{p^s}$ is an extension of $\mathbb{F}_{p^r}$, then $r \mid s$.

16.18 Verify the formula in Theorem 16.7 for $p = 2$ and $s = 4$ directly.

16.19 (a) Let F be a finite field. An element $\zeta \in F$ is an mth root of 1 if

$$\zeta^m = 1 .$$

Show that F contains m, mth roots of 1, in other words that $x^m - 1$ splits in F, if and only if

$$m \mid (|F| - 1) .$$

(b) Show that the splitting field of $x^m - 1 \in \mathbb{F}_p[x]$ is $\mathbb{F}_{p^r}$, where r is the smallest number such that

$$m \mid (p^r - 1) .$$

16.20 The Fibonacci numbers $f(n)$, are given by the recursion formula

$$f(n + 1) = f(n) + f(n - 1), \quad f(0) = 0, \ f(1) = 1 .$$

In matrix notation this may be written

$$\begin{pmatrix} f(n + 1) \\ f(n) \end{pmatrix} = \begin{pmatrix} 1 & 1 \\ 1 & 0 \end{pmatrix} \begin{pmatrix} f(n) \\ f(n - 1) \end{pmatrix}, \quad \begin{pmatrix} f(1) \\ f(0) \end{pmatrix} = \begin{pmatrix} 1 \\ 0 \end{pmatrix} .$$

This definition makes sense in $\mathbb{F}_p$. Compute the Fibonacci numbers mod p for a few primes p. Notice that they seem to be periodic. They can also be given by the closed formula

$$f(n) = \frac{\zeta^n - (-\zeta)^{-n}}{\zeta + \zeta^{-1}} ,$$

where $\zeta^2 - \zeta - 1 = 0$. This holds over the integers and the integers mod p. Explain why the Fibonacci numbers mod p are periodic. Determine their period. (*Mathematica* has a function `Fibonacci` which you may find useful. You can produce a list of the first n Fibonacci numbers mod p with `Mod[ Table[ Fibonacci[k], {k,1,n} ], p ]`).

Chapter 17

Galois Groups

Introduction

In Chapter 7, we looked at some examples of symmetry groups of equations. A symmetry of an equation was defined as a permutation of its roots that preserved any algebraic relations among them. We were not very exact about what we meant by "algebraic relations among the roots." In the last chapter, we made this more precise and saw that the structure of the splitting field of an equation reflects these algebraic relations. So a natural way to define what a symmetry of an equation should be is to say that it should be a mapping of the splitting field to itself that preserves the structure of the field and fixes the coefficients of the equation. As we shall show shortly, this implies that it will permute the roots. To put it more succinctly: if we have a polynomial f with coefficients in a field F, with a splitting field E/F, then a symmetry is an automorphism of E that fixes F.

Let us look at the examples from Chapter 7 again from this point of view. Take

$$f(x) = x^4 + x^3 + x^2 + x + 1 .$$

As we saw in Example 16(ii) the splitting field of f is $\mathbb{Q}(\omega)$, where ω is a primitive 5th root of 1. The other roots are ω^2, ω^3, and ω^4. Suppose that α is an automorphism of $\mathbb{Q}(\omega)$. Since every element of this field can be written as a polynomial in ω, $\alpha(\omega)$ determines $\alpha(\zeta)$ for any ζ in the field. So what can $\alpha(\omega)$ be? Notice that if ζ is a root of f, then

$$0 = \alpha(\zeta^4 + \zeta^3 + \zeta^2 + \zeta + 1) = \alpha(\zeta)^4 + \alpha(\zeta)^3 + \alpha(\zeta)^2 + \alpha(\zeta) + 1 .$$

So $\alpha(\zeta)$ is also a root of f. Therefore $\alpha(\omega)$ must be one of ω, ω^2, ω^3, or ω^4. Define automorphisms α_j by

$$\alpha_j(\omega) = \omega^j ,$$

for $j = 1, 2, 3, 4$. Then as we saw in Chapter 7,

$$\alpha_2^2 = \alpha_4 , \quad \alpha_2^3 = \alpha_3 , \quad \alpha_2^4 = \alpha_1 = 1 .$$

Thus the group of automorphisms of the splitting field is cyclic of order 4.

Next take

$$f(x) = x^4 - 10x^2 + 1 \in \mathbb{Q}[x] \,.$$

In Example 16.3 we saw that the splitting field of f is $\mathbb{Q}(\sqrt{2}, \sqrt{3})$. Since any element of this field can be expressed in terms of $\sqrt{2}$ and $\sqrt{3}$, an automorphism α is determined by $\alpha(\sqrt{2})$ and $\alpha(\sqrt{3})$. Now

$$0 = \alpha\big((\sqrt{2})^2 - 2\big) = \alpha(\sqrt{2})^2 - 2 \,.$$

Therefore

$$\alpha(\sqrt{2}) = \pm\sqrt{2} \,.$$

Similarly, we must have that

$$\alpha(\sqrt{3}) = \pm\sqrt{3} \,.$$

So there are four possibilities for α:

α	$\alpha(\sqrt{2})$	$\alpha(\sqrt{3})$
α_1	$\sqrt{2}$	$\sqrt{3}$
α_2	$-\sqrt{2}$	$\sqrt{3}$
α_3	$\sqrt{2}$	$-\sqrt{3}$
α_4	$-\sqrt{2}$	$-\sqrt{3}$.

We see that

$$\alpha_1 = 1 \quad \text{and} \quad \alpha_4 = \alpha_2\alpha_3 \,.$$

Thus, the group of symmetries is isomorphic to V, as we saw in Chapter 7.

Let us do one more example:

Example 17.1

Set $f(x) = x^4 - 2x^2 - 2 \in \mathbb{Q}[x]$ (see Exercise 7.12). In Exercise 16.7, you saw that the roots of f are

$$\zeta_1 = \sqrt{1 + \sqrt{3}} \qquad \zeta_2 = \sqrt{1 - \sqrt{3}}$$

$$\zeta_3 = -\zeta_1 = -\sqrt{1 + \sqrt{3}} \qquad \zeta_4 = -\zeta_2 = -\sqrt{1 - \sqrt{3}} \,,$$

and that the splitting field is

$$E = \mathbb{Q}\,(\zeta_1, \zeta_2) \,.$$

Now ζ_1 and ζ_2 satisfy the relation

$$\zeta_1^2 + \zeta_2^2 = 2 \,.$$

In terms of E, this means that E has two subfields, $\mathbb{Q}(\zeta_1)$ and $\mathbb{Q}(\zeta_2)$, with

$$\mathbb{Q}(\zeta_1) \cap \mathbb{Q}(\zeta_2) = \mathbb{Q}(\zeta_1^2) = \mathbb{Q}(\zeta_2^2) = \mathbb{Q}(\sqrt{3}) \,.$$

Here is a diagram of the extensions:

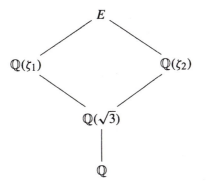

$$(17.1)$$

Any automorphism α of E is determined by $\alpha(\zeta_1)$ and $\alpha(\zeta_2)$, and must satisfy

$$\alpha(\zeta_1)^2 + \alpha(\zeta_2)^2 = 2 \,.$$

This gives us eight possibilities for α. The last column gives the corresponding permutation of the four roots.

α	$\alpha(\zeta_1)$	$\alpha(\zeta_2)$	permutation
α_1	ζ_1	ζ_2	(1)
α_2	$-\zeta_1$	ζ_2	(1 3)
α_3	ζ_1	$-\zeta_2$	(2 4)
α_4	$-\zeta_1$	$-\zeta_2$	(1 3)(2 4)
α_5	ζ_2	ζ_1	(1 2)(3 4)
α_6	$-\zeta_2$	ζ_1	(1 4 3 2)
α_7	ζ_2	$-\zeta_1$	(1 2 3 4)
α_8	$-\zeta_2$	$-\zeta_1$	(1 4)(2 3)

The list of eight permutations on the right is the list in the table at the beginning of Chapter 3! So the group of automorphisms of E is isomorphic to D_4. ⬚

REMARK 17.1   We have been using the following principle. Suppose that $f \in F[x]$, and E is an extension of F containing a root ζ of f. Let α be an automorphism of E that fixes F. Then $\alpha(\zeta)$ is also a root of f. Why is this so? Write f out:

$$f(x) = a_n x^n + \cdots + a_1 x + a_0 \,,$$

where $a_n, \ldots, a_1, a_0 \in F$. So

$$0 = a_n \zeta^n + \cdots + a_1 \zeta + a_0 \,.$$

Applying α to this equation gives us

$$
\begin{aligned}
0 &= \alpha \left(a_n \zeta^n + \cdots + a_1 \zeta + a_0 \right) \\
&= \alpha(a_n)\alpha(\zeta)^n + \cdots + \alpha(a_1)\alpha(\zeta) + \alpha(a_0) \\
&= a_n \alpha(\zeta)^n + \cdots + a_1 \alpha(\zeta) + \alpha(a_0) \\
&= f\big(\alpha(\zeta)\big) \, .
\end{aligned}
$$

Thus α permutes the roots of f.

Definition

We now are ready to make a formal definition of the group of symmetries of an equation. Following the discussion above, we shall define it as the group of automorphisms of its splitting field, which leave the field of coefficients invariant. In fact we can look at this group for any extension. If E/F is a field extension, then we say that an automorphism α of E *fixes* F if $\alpha(a) = a$ for all $a \in F$. The set of all automorphisms of E, which fixes F, forms a subgroup of the group of automorphisms of E.

DEFINITION 17.1   *Let E/F be a field extension. Then the Galois group of E/F, written $\mathrm{Gal}(E/F)$, is the group of automorphisms of E, which fix F, in other words*

$$\mathrm{Gal}(E/F) \;=\; \{\alpha \mid \alpha \text{ is an automorphism of E, } \alpha(a) = a \text{ for all } a \in F\} \, .$$

If E is the splitting field of a polynomial $f \in F[x]$, then $\mathrm{Gal}(E/F)$ is called the Galois group of f, and written $\mathrm{Gal}(f)$.

So for $f(x) = x^4 + x^3 + x^2 + x + 1$,

$$\mathrm{Gal}(f) = \mathrm{Gal}\left(\mathbb{Q}(\omega)/\mathbb{Q}\right) \cong \mathbb{Z}/4\mathbb{Z} \, ,$$

and for $f(x) = x^4 - 10x^2 + 1$,

$$\mathrm{Gal}(f) = \mathrm{Gal}\left(\mathbb{Q}(\sqrt{2}, \sqrt{3})/\mathbb{Q}\right) \cong V$$

Here is an example where E is not the splitting field of a polynomial: take $E = \mathbb{Q}(\sqrt[3]{2})$. Any automorphism of E must map $\sqrt[3]{2}$ to a cube root of 2. But as we saw in Example 16.2, the other two cube roots of 2 do not lie in E. Therefore

$$\mathrm{Gal}(E/\mathbb{Q}) = \{1\} \, .$$

As we noted in Exercise 14.10, if E is an extension of $\mathbb{Q}$, then every automorphism of E fixes $\mathbb{Q}$.

Example 17.1 Continued We computed that if $f(x) = x^4 - 2x^2 - 2$, then

$$\text{Gal}(f) = \text{Gal}(E/\mathbb{Q}) \cong D_4 .$$

We can also look at the extensions $E/\mathbb{Q}(\zeta_1)$, $E/\mathbb{Q}(\zeta_2)$, and $E/\mathbb{Q}(\sqrt{3})$, and compute their Galois groups. Every automorphism of E, which fixes $\mathbb{Q}(\zeta_1)$, $\mathbb{Q}(\zeta_2)$, or $\mathbb{Q}(\sqrt{3})$, fixes $\mathbb{Q}$, and therefore belongs to $\text{Gal}(E/\mathbb{Q})$. So $\text{Gal}\left(E/\mathbb{Q}(\zeta_1)\right)$, $\text{Gal}\left(E/\mathbb{Q}(\zeta_2)\right)$, and $\text{Gal}\left(E/\mathbb{Q}(\sqrt{3})\right)$ are subgroups of $\text{Gal}(E/\mathbb{Q})$. Looking at the table, we see that

$$\text{Gal}(E/\mathbb{Q}(\zeta_1)) = \langle \alpha_3 \rangle ,$$

which is cyclic of order 2. Similarly,

$$\text{Gal}(E/\mathbb{Q}(\zeta_2)) = \langle \alpha_2 \rangle .$$

Now elements of $\text{Gal}(E/\mathbb{Q}(\sqrt{3}))$ are automorphisms that fix ζ_1^2. The table tells us, then, that

$$\text{Gal}(E/\mathbb{Q}(\sqrt{3})) = \{\alpha_1, \alpha_2, \alpha_3, \alpha_4\} \cong V .$$

Here is the diagram showing the inclusions.

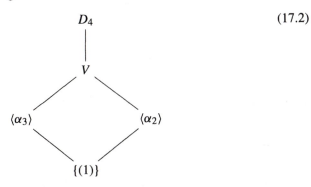

$$(17.2)$$

We have gone from diagram (17.1) to this one. You can also start from this diagram and construct diagram (17.1), as we shall soon see. □

Let us compute one example that is a little more sophisticated. We shall calculate the Galois group of the cyclotomic field $E_n/\mathbb{Q}$, the splitting field of Φ_n. In the introduction, we looked at the case $n = 5$. Pick a primitive nth root of 1, ω. So $E_n = \mathbb{Q}(\omega)$. Any element $\alpha \in \text{Gal}(\mathbb{Q}(\omega)/\mathbb{Q})$ is determined by $\alpha(\omega)$, which must be another primitive nth root of 1. Thus for each j, $(j, n) = 1$, $1 \le j < n$, there is an automorphism α_j, defined by

$$\alpha_j(\omega) = \omega^j .$$

For j, k relatively prime to n,

$$\alpha_j\left(\alpha_k(\omega)\right) = \left(\omega^k\right)^j = \omega^{jk} = \alpha_l(\omega) ,$$

where $1 \le l < n$, $l \equiv jk \pmod{n}$. Therefore we can define a homomorphism

$$h : (\mathbb{Z}/n\mathbb{Z})^\times \to \text{Gal}\left(E_n/\mathbb{Q}\right) = \text{Gal}\left(\Phi_n\right)$$

by

$$h(j) = \alpha_j .$$

Since the α_j's are all the automorphisms of E_n, h is surjective. For different j's the α_j's are different. So h is injective, and therefore an isomorphism. Recall that we discussed the structure of $(\mathbb{Z}/n\mathbb{Z})^\times$ in Chapter 13, pages 172-174. In particular, if n is prime, then $\text{Gal}(\Phi_n)$ is cyclic of order $n - 1$.

How Large is the Galois Group?

These calculations also suggest that there is a connection between $[E : F]$ and $|\text{Gal}(E/F)|$. In every example, we see that

$$|\text{Gal}(E/F)| \leq [E : F] ,$$

with equality when E is the splitting field of a polynomial $f \in F[x]$. This is in fact true if f has distinct roots. To prove it, we have to construct automorphisms of E that fix F. Such an automorphism can be regarded as an extension of the identity map on F to E. More generally, we can look at extending an arbitrary automorphism of F to E or extending an isomorphism between two fields.

THEOREM 17.1
Let F_1 and F_2 be fields, and

$$\psi : F_1 \rightarrow F_2$$

an isomorphism. Suppose that $f_1 \in F_1[x]$ is a monic polynomial (not constant), and $f_2 \in F_2[x]$ the corresponding polynomial under ψ. Let E_1/F_1 be the splitting field of f_1, and E_2/F_2, the splitting field of f_2.

Then ψ extends to an isomorphism $E_1 \rightarrow E_2$. The number of such extensions is at most $[E_1 : F_1]$ and is exactly $[E_1 : F_1]$ if the roots of f_1 in E_1 are distinct.

Before starting the proof, we should clarify some points. First, if

$$f_1(x) = a_n x^n + \cdots + a_1 x + a_0 ,$$

with $a_n, \ldots , a_1, a_0 \in F_1$, let

$$b_j = \psi(a_j) , \quad 0 \leq j \leq n .$$

Then

$$f_2(x) = \psi_*(f_1) := b_n x^n + \cdots + b_1 x + b_0 \in F_2[x] ,$$

as discussed in Exercise 14.12. Secondly, as indicated above, the inequality for $|\text{Gal}(E/F)|$ will follow from the special case where $F_1 = F_2 = F$, ψ is the identity on F, and $E_1 = E_2 = E$. But the proof of the theorem is by induction, and for the induction step, we need the more general result. Thirdly, if we take $F_1 = F_2 = F$ and ψ the identity on F, and let E_1 and E_2 be two splitting fields of f over F, then the theorem shows that E_1 and E_2 are isomorphic. In other words, splitting fields are unique up to isomorphism. We begin with a lemma:

LEMMA 17.1
Suppose that f_1 (and therefore f_2) is irreducible. Let ζ_1 be a root of f_1 in E_1. Then ψ can be extended to a homomorphism

$$\chi : F_1(\zeta_1) \to E_2 ,$$

and the number of such extensions is the number of distinct roots of f_2 in E_2.

PROOF From Exercise 14.12, we have a ring isomorphism

$$\psi_* : F_1[x] \xrightarrow{\ \cong\ } F_2[x] ,$$

which induces an isomorphism of fields

$$\bar{\psi}_* : F_1[x]/(f_1) \xrightarrow{\ \cong\ } F_2[x]/(f_2) .$$

Let ζ_2 be a root of f_2 in E_2. As we saw in Remark 14.4, evaluation at ζ_1 and ζ_2 define isomorphisms:

$$\bar{\epsilon}_{\zeta_1} : F_1[x]/(f_1) \xrightarrow{\ \cong\ } F_1(\zeta_1) , \quad \bar{\epsilon}_{\zeta_2} : F_2[x]/(f_2) \xrightarrow{\ \cong\ } F_2(\zeta_2) .$$

Putting all these together, we get a homomorphism

$$\chi : F_1(\zeta_1) \xrightarrow{\ \bar{\epsilon}_{\zeta_1}^{-1}\ } F_1[x]/(f_1) \xrightarrow{\ \bar{\psi}_*\ } F_2[x]/(f_2) \xrightarrow{\ \bar{\epsilon}_{\zeta_2}\ } F_2(\zeta_2) \hookrightarrow E_2$$

which extends $\psi : F_1 \to F_2$. Different roots ζ_2 in E_2 give different homomorphisms.

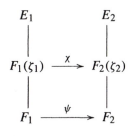

The theorem is proved by adjoining roots of f_1 in steps and extending ψ step by step, as in the proof of the existence of splitting fields.

PROOF We argue by induction on $n := [E_1 : F_1]$. If f_1 is linear, then there is nothing to prove. Assume that the result holds for splitting fields of degree less than

n. We want to use the lemma to extend ψ to $F_1(\zeta_1)$, for some root ζ_1 of f_1. Then we can use the induction assumption to extend further to E_1. However, the lemma requires that f_1 be irreducible. So let g_1 be an irreducible factor of degree > 1 of f_1, and let g_2 be the corresponding factor of f_2. Pick a root ζ_1 of g_1 in E_1. As shown in the lemma we can extend ψ to a homomorphism from $F_1(\zeta_1)$ into E_2, and the number of such homomorphisms is the number of distinct roots of g_2 in E_2. As g_1 splits in E_1, this is at most

$$\deg g_2 = \deg g_1 = [F_1(\zeta_1) : F_1] \ ,$$

and is equal to $[F_1(\zeta_1) : F_1]$ if the roots are distinct.

Let $\tilde{\psi}$ be one such homomorphism. Set $\tilde{F}_1 = F_1(\zeta_1)$, and $\tilde{F}_2 = \tilde{\psi}(F_1(\zeta_1))$:

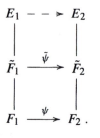

Now regard f_1 as a polynomial in $\tilde{F}_1[x]$. Then E_1 is a splitting field for f_1 over $\tilde{F}_1$, and E_2 a splitting field for f_2 over $\tilde{F}_2$. As $[E_1 : \tilde{F}_1] < [E_1 : F_1]$, the induction assumption applies. So $\tilde{\psi}$ extends to an isomorphism $E_1 \rightarrow E_2$. Furthermore, the number of such extensions is at most $[E_1 : \tilde{F}_1]$, and is equal to $[E_1 : \tilde{F}_1]$ if the roots of f_1 are distinct.

Now we can count the number of extensions of ψ to E_1. There are at most $[\tilde{F}_1 : F_1]$ extensions to $\tilde{F}_1$ and for each of these, at most $[E_1 : \tilde{F}_1]$ extensions from $\tilde{F}_1$ to E_1. So there are at most

$$\left[\tilde{F}_1 : F_1\right]\left[E_1 : \tilde{F}_1\right] = \left[E_1 : F_1\right]$$

extensions from F_1 to E_1. And if the roots of f_1 are distinct, then equality holds everywhere. This completes the induction step. Therefore by the principle of induction, the theorem is proved. ∎

COROLLARY 17.1

Let F be a field and f a polynomial with coefficients in F. Suppose that E_1 and E_2 are splitting fields for f over F. Then $E_1 \cong E_2$; in fact there is an isomorphism from E_1 to E_2 that fixes F.

PROOF   Take $F_1 = F_2 = F$ and $\psi = 1_F$ in the theorem above. It tells us then that there exists an isomorphism from E_1 to E_2 that fixes F. ∎

COROLLARY 17.2

Suppose E is the splitting field of a polynomial $f \in F[x]$. Then

$$|\mathrm{Gal}(E/F)| \le [E:F] .$$

If the roots of f are distinct, then

$$|\mathrm{Gal}(E/F)| = [E:F] .$$

PROOF Take $F_1 = F_2 = F$, $E_1 = E_2 = E$, and $\psi = 1_F$ in the theorem. Then extensions of ψ to E are just automorphisms of E that fix F, in other words, elements of the Galois group. ∎

These results raise the question: when are the roots of an irreducible polynomial distinct? Exercise 14.20 shows that a polynomial $f \in F[x]$ has a repeated factor in $F[x]$ if and only if

$$(f, f') \ne 1 .$$

If E/F is a splitting field for f, we can regard f as a polynomial in $E[x]$. Then the same criterion tells us whether f has a multiple root. We can apply this to answer our question.

The answer depends on whether the characteristic of F is positive or not because in positive characteristic, a polynomial can have derivative 0 without being constant. So to begin with, assume that f is irreducible and not constant and that char $F = 0$. Then

$$\deg f' < \deg f .$$

But if $(f, f') \ne 1$, then

$$(f, f') = f$$

since f is irreducible. It follows that $f' = 0$. This implies that f would have to be constant. So in characteristic 0, an irreducible polynomial cannot have multiple roots.

If char $F = p$, then it can happen that an irreducible polynomial, which is not constant, has a multiple root. However, this does not happen if F is finite.

THEOREM 17.2

Let F be a finite field. Then an irreducible polynomial $f \in F[x]$ does not have repeated roots in its splitting field.

PROOF Suppose that char $F = p$ and that f has a multiple root in its splitting field. Then applying the criterion as before, we see that

$$f' = 0 .$$

Write out f:

$$f(x) = a_n x^n + \cdots + a_1 x + a_0 ,$$

for some $n > 0$, and some $a_0, a_1, \ldots, a_n \in F$. Therefore

$$0 = f'(x) = na_n x^{n-1} + \cdots + a_1 .$$

So for all $j > 0$,

$$ja_j = 0 ,$$

which means that if $a_j \neq 0$, then $j \equiv 0 \pmod{p}$. Thus f is of the form

$$f(x) = a_{kp} x^{kp} + \cdots + a_p x^p + a_0 .$$

Now Exercise 14.22 tells that since F is finite, every element of F is a pth power. Therefore for every $j \leq k$, there exists a $b_{jp} \in F$ such that

$$a_{jp} = b_{jp}^p .$$

This gives us

$$f(x) = b_{kp}^p x^{kp} + \cdots + b_p^p x^p + b_0^p = \left(b_k x^k + \cdots + b_p x + b_0\right)^p$$

(see Exercise 14.22). But f is irreducible. So this is impossible. Therefore f does not have a repeated root. ∎

This discussion is worth summarizing. First we create a definition.

DEFINITION 17.2   *A polynomial $f \in F[x]$ is separable if none of its irreducible factors has a repeated root.*

THEOREM 17.3
Let F be a field of characteristic 0 or a finite field. Then every polynomial in $F[x]$ is separable.

The Galois Correspondence

As the earlier examples suggest, if we have a field extension E/F and an intermediate field K, $E \supset K \supset F$, then

$$\mathrm{Gal}(E/K) < \mathrm{Gal}(E/F)$$

since an automorphism of E, which fixes K, fixes F. On the other hand, if $H < \mathrm{Gal}(E/F)$, then it is easy to see that

$$\mathrm{Fix}(H) := \{a \in E \mid \alpha(a) = a, \text{ for all } \alpha \in H\} ,$$

is a subfield of E containing F. For a subgroup H and an intermediate field K, there are the obvious relations:

$$\text{Gal}\left(E/\text{Fix}(H)\right) \supset H$$
$$\text{Fix}\left(\text{Gal}(E/K)\right) \supset K \ .$$

Example 17.1 Concluded Let us continue the discussion of Example 17.1. We can compute $\text{Fix}(H)$ for the subgroups of $\text{Gal}(E/\mathbb{Q})$ in Diagram (17.2). First, we need one more simple remark. Suppose E/F is the splitting field of a separable polynomial $f \in F[x]$, and K is an intermediate field. Then E is also the splitting field of f regarded as a polynomial over K. Therefore by Corollary 17.2, $|\text{Gal}(E/K)| = [E : K]$.

Now let us begin our calculations. Clearly, for $H = \text{Gal}(E/\mathbb{Q})$ itself, $\text{Fix}(H) = \mathbb{Q}$. Next, take $H = \langle \alpha_2, \alpha_3 \rangle \cong V$. We already know that α_2 and α_3 fix $\mathbb{Q}(\sqrt{3})$. So

$$\text{Fix}(H) \supset \mathbb{Q}(\sqrt{3}) \ ,$$

and

$$[E : \text{Fix}(H)] \leq [E : \mathbb{Q}(\sqrt{3})] = 4 \ .$$

Therefore by the remark,

$$|\text{Gal}\left(E/\text{Fix}(H)\right)| = [E : \text{Fix}(H)] \leq 4 \ .$$

But

$$\text{Gal}\left(E/\text{Fix}(H)\right) \supset H,$$

which has order 4. So

$$4 = |\text{Gal}\left(E/\text{Fix}(H)\right)| = [E : \text{Fix}(H)] = [E : \mathbb{Q}(\sqrt{3})] \ ,$$

and

$$\text{Fix}(H) = \mathbb{Q}(\sqrt{3}) \ .$$

We will now take care of the two subgroups of order 2. The automorphism α_3 fixes ζ_1. We know that

$$|\langle \alpha_3 \rangle| = 2 = [E : \mathbb{Q}(\zeta_1)] \ ,$$

and applying the argument used for V, we find that

$$\text{Fix}(\langle \alpha_3 \rangle) = \mathbb{Q}(\zeta_1) \ .$$

Similarly, α_2 fixes ζ_2, and

$$\text{Fix}(\langle \alpha_2 \rangle) = \mathbb{Q}(\zeta_2) \ .$$

Lastly, for $H = \{(1)\}$, $\text{Fix}(H) = E$. So we have recovered Diagram (17.1) from the Lattice (17.2).

This discussion can be extended to all the subgroups of $\text{Gal}(E/\mathbb{Q})$ since we know what the subgroups of D_4 are. Here is the subgroup lattice of $\text{Gal}(E/\mathbb{Q})$ (see Exercise 10.3). Diagram (17.2) is embedded on the right side.

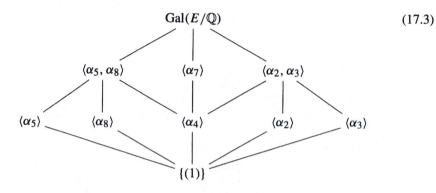 (17.3)

We can compute $\text{Fix}(H)$ for the other two subgroups of order 4 in the same way as we did for V. The subgroup $\langle \alpha_5, \alpha_8 \rangle$ fixes

$$\zeta_1 \zeta_2 = \sqrt{1 + \sqrt{3}}\sqrt{1 - \sqrt{3}} = \sqrt{-2}.$$

Since

$$[E : \mathbb{Q}(\sqrt{-2})] = 4,$$

the discussion above shows that

$$\text{Fix}(\langle \alpha_5, \alpha_8 \rangle) = \mathbb{Q}(\sqrt{-2}).$$

The automorphism α_7 fixes

$$(\zeta_1 \zeta_2)(\zeta_1^2 - 1) = \sqrt{-2}\sqrt{3} = \sqrt{-6}.$$

Therefore

$$\text{Fix}(\langle \alpha_7 \rangle) = \mathbb{Q}(\sqrt{-6}).$$

For the remaining subgroups of order 2, notice that α_5 fixes $\zeta_1 + \zeta_2$, and α_8 fixes $\zeta_1 - \zeta_2$. Since

$$\langle \alpha_4 \rangle = \langle \alpha_2, \alpha_3 \rangle \cap \langle \alpha_5, \alpha_8 \rangle,$$

α_4 fixes both $\sqrt{3}$ and $\sqrt{-2}$. Therefore, we have

$$\text{Fix}(\langle \alpha_5 \rangle) = \mathbb{Q}(\zeta_1 + \zeta_2)$$
$$\text{Fix}(\langle \alpha_8 \rangle) = \mathbb{Q}(\zeta_1 - \zeta_2)$$
$$\text{Fix}(\langle \alpha_4 \rangle) = \mathbb{Q}(\sqrt{3}, \sqrt{-2}).$$

We can put these fields in a diagram that extends Diagram (17.1):

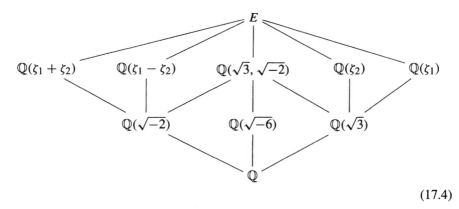

$$(17.4)$$

This is just the lattice of subgroups (17.3) turned upside down!   ⬚

In our discussion, we have found that for each subgroup H,

$$\text{Gal}\left(\text{Fix}(H)\right) = H .$$

It is easy to see that for any intermediate field,

$$\text{Fix}\left(\text{Gal}(E/K)\right) = K .$$

These two statements imply that there is a one-to-one correspondence between subgroups and intermediate fields. This is the *Galois correspondence*. The key to proving these results in general is the following theorem.

THEOREM 17.4
Let E be a field and G a finite group of automorphisms of E. Set $F = \text{Fix}(G)$. Then

$$[E : F] \leq |G| .$$

PROOF   Set $m = |G|$. What we shall do is to show that any n elements of E are linearly dependent over F, if $n > m$. First list the elements of G:

$$G = \{\alpha_1 = 1, \alpha_2, \ldots, \alpha_m\} .$$

Pick n elements $\zeta_1, \ldots, \zeta_n \in E$, and consider the system of m equations in n unknowns in E:

$$\begin{pmatrix} \alpha_1(\zeta_1) & \cdots & \alpha_1(\zeta_n) \\ \vdots & & \vdots \\ \alpha_m(\zeta_1) & \cdots & \alpha_m(\zeta_n) \end{pmatrix} \begin{pmatrix} x_1 \\ \vdots \\ x_n \end{pmatrix} = 0 .$$

Notice that since $\alpha_1 = 1$, the first equation is just

$$x_1\zeta_1 + \cdots + x_n\zeta_n = 0 .$$

What we want is a solution of this equation that lies in F. Now since the number of unknowns exceeds the number of equations, there exists a nontrivial solution in E. Pick one with as few nonzero entries as possible, say $(\xi_1, \dots, \xi_n)$. We can reorder the entries so that $\xi_1 \neq 0$. Since the system is linear,

$$\xi_1^{-1}(\xi_1, \dots, \xi_n)$$

is also a solution, with the minimal number of nonzero entries. So we can assume that $\xi_1 = 1$. The system then looks like this:

$$\begin{pmatrix} \alpha_1(\zeta_1) & \cdots & \alpha_1(\zeta_n) \\ \alpha_2(\zeta_1) & \cdots & \alpha_2(\zeta_n) \\ \vdots & & \vdots \\ \alpha_m(\zeta_1) & \cdots & \alpha_m(\zeta_n) \end{pmatrix} \begin{pmatrix} 1 \\ \xi_2 \\ \vdots \\ \xi_n \end{pmatrix} = 0 .$$

What we want to do is to show that this solution lies in F, in other words, that it is invariant under G.

Suppose it is not. Then one of the nonzero entries is not invariant, say ξ_2. This means that there exists an element $\alpha_j \in G$ such that

$$\alpha_j(\xi_2) \neq \xi_2 .$$

Apply α_j to the system:

$$\begin{pmatrix} \alpha_j\alpha_1(\zeta_1) & \cdots & \alpha_j\alpha_1(\zeta_n) \\ \alpha_j\alpha_2(\zeta_1) & \cdots & \alpha_j\alpha_2(\zeta_n) \\ \vdots & & \vdots \\ \alpha_j\alpha_m(\zeta_1) & \cdots & \alpha_j\alpha_m(\zeta_n) \end{pmatrix} \begin{pmatrix} 1 \\ \alpha_j(\xi_2) \\ \vdots \\ \alpha_j(\xi_n) \end{pmatrix} = 0 .$$

Since G is a group,

$$\{\alpha_j\alpha_1, \alpha_j\alpha_2, \dots, \alpha_j\alpha_m\} = \{\alpha_1, \alpha_2, \dots, \alpha_m\} .$$

So this system is just the original one with the equations permuted. Therefore $(1, \alpha_j(\xi_2), \dots, \alpha_j(\xi_n))$ is also a solution of our system. It follows that

$$\left(1, \alpha_j(\xi_2), \dots, \alpha_j(\xi_n)\right) - \left(1, \xi_2, \dots, \xi_n\right) = \left(0, \alpha_j(\xi_2) - \xi_2, \dots, \alpha_j(\xi_n) - \xi_n\right)$$

is a solution. It is nontrivial since the second entry is not 0. But it has one less nonzero entry than the original solution. This is impossible. Therefore the original solution is invariant under G. In other words, $\{\zeta_1, \dots, \zeta_n\}$ is linearly dependent over F. This proves that

$$[E : F] \leq |G| . \quad \blacksquare$$

We can now prove the fundamental theorem about the Galois correspondence. It will allow us to translate questions about polynomials and their roots into questions about their Galois groups.

THEOREM 17.5 (Fundamental Theorem of Galois Theory)
Let E/F be the splitting field of a separable polynomial in $F[x]$. Then there is a one-to-one correspondence between subgroups of $\mathrm{Gal}(E/F)$ and intermediate fields between E and F given by

$$H \leftrightarrow \mathrm{Fix}(H) \quad \text{and} \quad \mathrm{Gal}(E/K) \leftrightarrow K \ ,$$

where H is a subgroup and K an intermediate field, such that

$$\mathrm{Gal}\left(E/\mathrm{Fix}(H)\right) = H \quad \text{and} \quad \mathrm{Fix}\left(\mathrm{Gal}(E/K)\right) = K \ .$$

PROOF Suppose that E/F is the splitting field of $f \in F[x]$, which is separable. First we show that

$$\mathrm{Gal}\left(E/\mathrm{Fix}(H)\right) = H \ ,$$

for any subgroup H. Now by Corollary 17.2

$$|\mathrm{Gal}\left(E/\mathrm{Fix}(H)\right)| = [E : \mathrm{Fix}(H)] \ ,$$

since E is the splitting field for f over $\mathrm{Fix}(H)$. And by Theorem 17.4, we know that

$$[E : \mathrm{Fix}(H)] \leq |H| \ .$$

Therefore

$$|\mathrm{Gal}\left(E/\mathrm{Fix}(H)\right)| \leq |H| \ .$$

On the other hand, as we remarked earlier,

$$\mathrm{Gal}\left(E/\mathrm{Fix}(H)\right) \supset H \ ,$$

so that

$$|\mathrm{Gal}\left(E/\mathrm{Fix}(H)\right)| \geq |H| \ .$$

Hence $|\mathrm{Gal}\left(E/\mathrm{Fix}(H)\right)| = |H|$, and

$$\mathrm{Gal}\left(E/\mathrm{Fix}(H)\right) = H \ .$$

Next we prove that

$$\mathrm{Fix}\left(\mathrm{Gal}(E/K)\right) = K \ .$$

Set $H := \mathrm{Gal}(E/K)$, and $K' := \mathrm{Fix}(H) \supset K$. We have just shown that

$$\mathrm{Gal}(E/K') = H = \mathrm{Gal}(E/K) \ .$$

Again, since E is the splitting field for f regarded as a polynomial over K or K',

$$[E : K] = |\mathrm{Gal}(E/K)| = |\mathrm{Gal}(E/K')| = [E : K'] \ .$$

But since $K' \supset K$,

$$[E : K] = [E : K'][K' : K] \ ,$$

and therefore $[K' : K] = 1$ and

$$\text{Fix}\left(\text{Gal}(E/K)\right) = K' = K . \quad \blacksquare$$

COROLLARY 17.3
$\text{Fix}\left(\text{Gal}(E/F)\right) = F.$

COROLLARY 17.4
Let H be a subgroup of $G =: \text{Gal}(E/F)$, and let $K = \text{Fix}(H)$. Then

$$[K : F] = [G : H] .$$

PROOF   By Theorem 17.5 we have

$$H = \text{Gal}(E/K) ,$$

and by Corollary 17.2

$$[E : K] = |\text{Gal}(E/K)| .$$

Therefore

$$[K : F] = [E : F]/[E : K] = |\text{Gal}(E/F)|/|\text{Gal}(E/K)| = [G : H]$$

Here is a diagram with the degrees of the extensions:

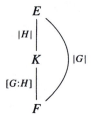

$\blacksquare$

Example 17.2
Let E be the splitting field of $x^p - 2$, p prime, as in Example 16.4. So $E = \mathbb{Q}(\sqrt[p]{2}, \omega)$, where ω is a primitive pth root of 1. Again, an automorphism α of E is determined by $\alpha(\sqrt[p]{2})$ and $\alpha(\omega)$. Now $\alpha(\sqrt[p]{2})$ must be another pth root of 2 by Remark 17.1. These are $\sqrt[p]{2}\,\omega^j$, where $0 \leq j < p$. And $\alpha(\omega)$ must be a primitive pth root of 1. So the automorphisms of E are $\alpha_{i,j}$, $1 \leq i < p, 0 \leq j < p$, where

$$\alpha_{i,j}(\omega) = \omega^i \qquad \text{and} \qquad \alpha_{i,j}\left(\sqrt[p]{2}\right) = \sqrt[p]{2}\,\omega^j .$$

To see what the group structure is, let us compute $\alpha_{i,j}\,\alpha_{k,l}$:

$$\alpha_{i,j}\big(\alpha_{k,l}(\omega)\big) = \alpha_{i,j}(\omega^k) = \omega^{ik}$$
$$\alpha_{i,j}\big(\alpha_{k,l}(\sqrt[p]{2}\,\omega)\big) = \alpha_{i,j}\big(\sqrt[p]{2}\,\omega^l\big) = \sqrt[p]{2}\,\omega^j\omega^{il} = \sqrt[p]{2}\,\omega^{j+il} \ .$$

Therefore

$$\alpha_{i,j}\alpha_{k,l} = \alpha_{ik,\,j+il} \ .$$

But this is just the group $F_{p(p-1)}$ as discussed in Exercise 8.22. So

$$\mathrm{Gal}(E/\mathbb{Q}) \cong F_{p(p-1)} \ .$$

It has the two interesting subgroups

$$T := \big\{\alpha_{1,j} \mid 0 \le j < p\big\} \cong \mathbb{F}_p$$

and

$$K := \big\{\alpha_{i,0} \mid 1 \le j < p\big\} \cong \mathbb{F}_p^{\times} \ .$$

Here is the lattice:

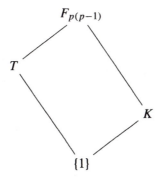

What are the corresponding intermediate fields between E and $\mathbb{Q}$? Well, T fixes $\mathbb{Q}(\omega)$ since

$$\alpha_{1,j}(\omega) = \omega \ ,$$

for all j. By the fundamental theorem,

$$[E : \mathrm{Fix}(T)] = |T| = p = [E : \mathbb{Q}(\omega)] \ .$$

Therefore

$$\mathrm{Fix}(T) = \mathbb{Q}(\omega) \ .$$

Similarly,

$$\mathrm{Fix}(K) = \mathbb{Q}(\sqrt[p]{2}) \ .$$

The diagram of field extensions is

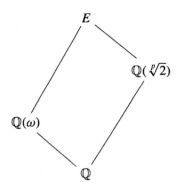

$$\square$$

We can also compute the Galois groups $\mathrm{Gal}(\mathbb{F}_{p^r}/\mathbb{F}_p)$:

THEOREM 17.6
The groups $\mathrm{Gal}(\mathbb{F}_{p^r}/\mathbb{F}_p)$ *are cyclic of order* r, *generated by the Frobenius homomorphism.*

PROOF   Let
$$\psi : \mathbb{F}_{p^r} \to \mathbb{F}_{p^r} ,$$
given by
$$\psi(a) = a^p ,$$
be the Frobenius homomorphism. As was shown in Exercise 14.22, it is an automorphism of $\mathbb{F}_{p^r}$, and it does fix $\mathbb{F}_p$ (see Exercise 14.10). So ψ belongs to $\mathrm{Gal}(\mathbb{F}_{p^r}/\mathbb{F}_p)$. Now
$$\psi^s(a) = a^{p^s} .$$
Suppose that $s \mid r$. We saw on page 237 that $\mathbb{F}_{p^s}$ is the splitting field of $x^{p^s} - x$ over $\mathbb{F}_p$. This means that $a^{p^s} = a$ for all $a \in \mathbb{F}_{p^s}$, but for $a \notin \mathbb{F}_{p^s}$, $a^{p^s} \neq a$. In other words,
$$\mathrm{Fix}\langle \psi^s \rangle = \mathbb{F}_{p^s} .$$
In particular for $s = r$, we have
$$|\psi| = r .$$
But
$$|\mathrm{Gal}\left(\mathbb{F}_{p^r}/\mathbb{F}_p\right)| = \left[\mathbb{F}_{p^r} : \mathbb{F}_p\right] = r .$$
Therefore
$$\mathrm{Gal}\left(\mathbb{F}_{p^r}/\mathbb{F}_p\right) = \langle \psi \rangle .$$

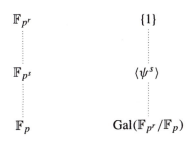

$\mathbb{F}_{p^r}$ $\{1\}$

$\mathbb{F}_{p^s}$ $\langle \psi^s \rangle$

$\mathbb{F}_p$ $\mathrm{Gal}(\mathbb{F}_{p^r}/\mathbb{F}_p)$

∎

The Galois correspondence reflects further properties of the structure of the Galois group.

THEOREM 17.7

(i) *Let K/F be the splitting field of a separable polynomial in $F[x]$, and let E/K be a field extension. Then $\mathrm{Gal}(E/K)$ is a normal subgroup of $\mathrm{Gal}(E/F)$. (This holds particularly if E/F is also the splitting field of a separable polynomial.)*

(ii) *Let E/F be the splitting field of a separable polynomial in $F[x]$, and set $G = \mathrm{Gal}(E/F)$. Suppose that H is a normal subgroup of G, and set $K = \mathrm{Fix}(H)$. Then*

$$\mathrm{Gal}(K/F) \cong G/H \ .$$

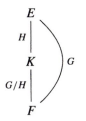

PROOF

(i) Suppose that K is the splitting field of $f \in F[x]$. Take $\alpha \in \mathrm{Gal}(E/F)$. By Remark 17.1, α permutes the roots of f. Therefore, for any root ζ of f, $\alpha(\zeta) \in K$. Since every element of K is a rational expression in the roots of f, it follows that $\alpha(K) \subset K$.

Now let $\beta \in \mathrm{Gal}(E/K)$ and $\zeta \in K$. Then $\alpha^{-1}(\zeta) \in K$. So $\beta(\alpha^{-1}(\zeta)) = \alpha^{-1}(\zeta)$. Therefore

$$(\alpha\beta\alpha^{-1})(\zeta) = \alpha(\alpha^{-1}(\zeta)) = \zeta \ .$$

Thus, $\alpha\beta\alpha^{-1} \in \mathrm{Gal}(E/K)$, and $\mathrm{Gal}(E/K)$ is a normal subgroup of $\mathrm{Gal}(E/F)$.

(ii) We first show that any $\alpha \in G$ maps $K = \text{Fix}(H)$ into itself. Take $\zeta \in K$ and $\beta \in H$. Then since H is normal,

$$\beta\alpha = \alpha\beta' \, ,$$

for some $\beta' \in H$, and

$$\beta\big(\alpha(\zeta)\big) = \alpha\big(\beta'(\zeta)\big) = \alpha(\zeta) \, .$$

Therefore $\alpha(\zeta) \in K$. Now we can define a homomorphism $\text{res} : G \to \text{Gal}(K/F)$ by setting

$$\text{res}(\alpha) = \alpha_{|K} \, ,$$

for $\alpha \in G$. Since $H = \text{Gal}(E/K)$, $H = \ker(\text{res})$.

It remains to show that res is surjective. This means that any $\beta \in \text{Gal}(K/F)$ should extend to an automorphism of E. Since E is the splitting field of a polynomial in $F[x]$, it is the splitting field of the same polynomial regarded as an element of $K[x]$. Therefore by Theorem 17.1, β extends to an automorphism of E, and res is surjective.

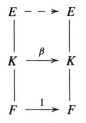

So res induces an isomorphism

$$G/H \xrightarrow{\cong} \text{Gal}(K/F) \, . \quad \blacksquare$$

In Example 17.1, all the subgroups of D_4 of order 4 are normal. For example take $H = \langle \alpha_2, \alpha_3 \rangle$. We calculated that $\text{Fix}(H) = \mathbb{Q}(\sqrt{3})$. Now

$$\alpha_5\left(\zeta_1^2\right) = \zeta_2^2 \, .$$

Since

$$\zeta_1^2 = 1 + \sqrt{3} \quad , \qquad \zeta_2^2 = 1 - \sqrt{3} \, ,$$

it follows that

$$\alpha_5(\sqrt{3}) = -\sqrt{3} \, .$$

Thus α_5 restricted to $\mathbb{Q}(\sqrt{3})$ generates $\text{Gal}\big(\mathbb{Q}(\sqrt{3})/\mathbb{Q}\big)$. And

$$\langle \alpha_5 \rangle \cong \text{Gal}(E/\mathbb{Q})\big/H \, .$$

Instead of α_5, we could have just as well used α_6, α_7, or α_8. The calculations for the other two groups of order 4 are similar.

There is only one normal subgroup of order 2:

$$\langle \alpha_4 \rangle = Z\big(\text{Gal}(E/\mathbb{Q})\big) \,.$$

We computed that $\text{Fix}(\langle \alpha_4 \rangle) = \mathbb{Q}(\sqrt{3}, \sqrt{-2})$. Now

$$\alpha_5 : \begin{cases} \sqrt{3} & \mapsto -\sqrt{3} \\ \sqrt{-2} & \mapsto \sqrt{-2} \end{cases} \qquad \alpha_2 : \begin{cases} \sqrt{3} & \mapsto \sqrt{3} \\ \sqrt{-2} & \mapsto -\sqrt{-2} \,. \end{cases}$$

So α_5 and α_2 restricted to $\mathbb{Q}(\sqrt{3}, \sqrt{-2})$ generate its automorphism group. And indeed,

$$\langle \alpha_5, \alpha_2 \rangle \cong \text{Gal}(E/\mathbb{Q})\big/\langle \alpha_4 \rangle \,.$$

Discriminants

In Chapter 15, the discriminant of a polynomial was introduced. In this section we discuss its connection with Galois theory.

Let F be a field, $f \in F[x]$ a separable polynomial of degree n, and E/F the splitting field of f. Denote by $\zeta_1, \ldots, \zeta_n$, the roots of f in E. Then

$$\Delta := \prod_{i<j} (\zeta_i - \zeta_j)^2 \in E$$

is the discriminant of f. As we saw, it is invariant under all permutations of the roots. In particular, it is fixed by $\text{Gal}(f) \subset S_n$. Therefore by Corollary 17.3,

$$\Delta \in F \,.$$

Now set

$$\delta = \prod_{i<j} (\zeta_i - \zeta_j) \in E \,.$$

So $\Delta = \delta^2$. From Exercise 15.11, we see that

$$\alpha(\delta) = (\text{sgn}\,\alpha)\delta \,,$$

for $\alpha \in S_n$. This means that δ is fixed by all even permutations, but not necessarily by all elements in $\text{Gal}(f)$. It is the invariant that determines whether $\text{Gal}(f) \subset A_n$ or not.

THEOREM 17.8
Suppose that f is a separable polynomial of degree n in $F[x]$. Then $\delta \in F$ if and only if $\text{Gal}(f) \subset A_n$.

PROOF Suppose that $\delta \in F$, and that $\alpha \in \mathrm{Gal}(f)$, but $\alpha \notin A_n$. Then α is an odd permutation of the roots and

$$\alpha(\delta) = -\delta \,.$$

This is impossible since $\delta \neq 0$. So $\mathrm{Gal}(f) \subset A_n$.

Conversely, if $\mathrm{Gal}(f) \subset A_n$, then as we saw above, δ is fixed by $\mathrm{Gal}(f)$. So $\delta \in F$ by Corollary 17.3. ∎

For example, if

$$f(x) = x^3 - x^2 - 2x + 1 \in \mathbb{Q}[x] \,,$$

then it is easy to check that f has no root in $\mathbb{Q}$ and is therefore irreducible. Its discriminant is 49. So the Galois group of f is A_3. In the next chapter, we will use δ in our analysis of quartics via Galois theory.

Exercises

17.1 The field $\mathbb{C}$ is the splitting field of $x^2 + 1 \in \mathbb{R}[x]$. Compute $\mathrm{Gal}(\mathbb{C}/\mathbb{R})$.

17.2 • Suppose that $f \in \mathbb{R}[x]$. By the fundamental theorem of algebra, we can assume that the splitting field of f is a subfield of $\mathbb{C}$. Show that complex conjugation defines an element of $\mathrm{Gal}(f)$.

17.3 Calculate the Galois group of $x^3 + 2x + 1 \in \mathbb{Q}[x]$.

17.4 Compute the Galois group of $x^4 - 4x^2 + 2 \in \mathbb{Q}[x]$.

17.5 Find the Galois group of $x^6 - 4x^3 + 1 \in \mathbb{Q}[x]$.

17.6 Let $h(x) = x^p - x - a \in \mathbb{F}_p[x]$, and let E be its splitting field. What is $[E : \mathbb{F}_p]$? Calculate $\mathrm{Gal}(h)$.

17.7 • Let F be a field of characteristic different from 2 or 3, and let $f \in F[x]$ be an irreducible cubic. In Chapter 16, page 232, it was shown that the splitting field E of f is of degree 3 over $F(\delta)$. Prove that

$$\mathrm{Gal}\left(E/F(\delta)\right) \cong A_3 \,.$$

17.8 • Let $f \in F[x]$ be an irreducible polynomial with distinct roots $\zeta_1, \ldots, \zeta_n$. Prove that the Galois group of f acts transitively on $\{\zeta_1, \ldots, \zeta_n\}$.

17.9 Suppose that E/F is a simple extension. Prove that if

$$|\mathrm{Gal}(E/F)| = [E : F]$$

then E is the splitting field of a separable polynomial in $F[x]$.

17.10 • Let ω be a primitive nth root of 1. Let $\theta := \omega + \omega^{-1}$.

    (a) Show that $[\mathbb{Q}(\omega) : \mathbb{Q}(\theta)] = 2$;

    (b) Identify $\mathrm{Gal}\left(\mathbb{Q}(\omega)/\mathbb{Q}(\theta)\right)$ as a subgroup of $\mathrm{Gal}\left(\mathbb{Q}(\omega)/\mathbb{Q}\right)$.

17.11 Suppose that F is a field of characteristic 0 containing the nth roots of unity. Let $a \in F$. Verify that $F(\sqrt[n]{a})$ is the splitting field of $x^n - a$ over F. Prove that $\mathrm{Gal}\left(F(\sqrt[n]{a})/F\right)$ is cyclic of order n.

17.12 Let E/F be a field extension, and $H < \mathrm{Gal}(E/F)$. Verify that

$$\mathrm{Fix}(H) := \{a \in E \mid \alpha(a) = a, \text{ for all } \alpha \in H\},$$

is a subfield of E containing F.

17.13 Verify the Galois correspondence explicitly for $\mathbb{Q}(\sqrt{2}, \sqrt{3})$, the splitting field for $x^4 - 10x^2 + 1 \in \mathbb{Q}[x]$, and its Galois group. In other words, write down the lattice of subfields of $\mathbb{Q}(\sqrt{2}, \sqrt{3})$ and the lattice of subgroups of its Galois group, and show which intermediate field corresponds to which subgroup.

17.14 Verify the Galois correspondence explicitly for the splitting field of $x^3 - 2 \in \mathbb{Q}[x]$ and its Galois group.

17.15 Compute the Galois group of $x^4 - 2 \in \mathbb{Q}[x]$ and verify the Galois correspondence.

17.16 Let $\bar{\mathbb{Q}} \subset \mathbb{C}$ be the field of algebraic numbers (see Exercise 16.8), and let $G = \mathrm{Gal}(\bar{\mathbb{Q}}/\mathbb{Q})$. For any $f \in \mathbb{Q}[x]$, show that there is a surjective homomorphism
$$G \rightarrow \mathrm{Gal}(f).$$

17.17 Let f be a separable polynomial in $F[x]$, and let E be its splitting field. Suppose that $H < \mathrm{Gal}(f)$, with $K = \mathrm{Fix}(H)$. Prove that for $\alpha \in \mathrm{Gal}(f)$, the field
$$\alpha K = \mathrm{Fix}\left(\alpha H \alpha^{-1}\right).$$

If $K = F(\eta)$, for some $\eta \in E$, show that
$$F(\alpha \eta) = \mathrm{Fix}\left(\alpha H \alpha^{-1}\right).$$

Chapter 18

Quartics

Galois Groups of Quartics

In this chapter, we look at what Galois theory says about quartic equations. First let us recall what we know about cubics. According to Exercise 17.8, the Galois group of an irreducible polynomial acts transitively on its roots. The only transitive subgroups of S_3 are A_3 and S_3 itself. So these are the only possible Galois groups if the cubic f is irreducible. (What if f is reducible?) Furthermore, its Galois group is A_3 if and only if the discriminant Δ is a square in F.

If

$$f(x) = x^3 + a_2 x^2 + a_1 x + a_0 , \quad a_2, a_1, a_0 \in F ,$$

and char $F \neq 3$, then we saw in Exercises 16.10 and 16.11 that the substitution $x = y - a_2/3$ transforms f into a cubic of the form

$$g(y) = y^3 + b_1 y + b_0 ,$$

where

$$b_1 = a_1 - \frac{a_2^2}{3} , \quad b_0 = a_0 - \frac{a_1 a_2}{3} + \frac{2 a_2^3}{27} ,$$

and the discriminant is

$$\Delta = -4 b_1^3 - 27 b_0^2 .$$

In terms of a_2, a_1, a_0, we have

$$\Delta = -27 a_0^2 - 4 a_1^3 + 18 a_0 a_1 a_2 + a_1^2 a_2^2 - 4 a_0 a_2^3 . \tag{18.1}$$

Now suppose that $f \in F[x]$ is an irreducible quartic with distinct roots $\zeta_1, \zeta_2, \zeta_3, \zeta_4$ in its splitting field E. Then $\mathrm{Gal}(f)$, which we will denote by G, can be identified with a subgroup of S_4, the group of permutations of the roots. We saw in Exercise 8.17 that the transitive subgroups of S_4 are

(i) S_4;

(ii) A_4;

(iii) $D_4 = \{(1\,2\,3\,4), (1\,3)(2\,4), (1\,4\,3\,2), (1), (2\,4), (1\,3), (1\,2)(3\,4), (1\,4)(2\,3)\}$
and its 3 conjugates;

(iv) $V = \{(1\,2)(3\,4), (1\,3)(2\,4), (1\,4)(2\,3), (1)\}$;

(v) $C_4 := \{(1\,2\,3\,4), (1\,3)(2\,4), (1\,4\,3\,2), (1)\}$ and its 3 conjugates.

So these are the possibilities for the Galois group of f. We need invariants that will help us decide which one of these it is.

Recall that a composition series for S_4 is given by

$$S_4 \rhd A_4 \rhd V \rhd \mathbb{Z}/2\mathbb{Z} \rhd \{1\}$$

(see (12.1)). We know that δ or Δ will tell us whether G is a subgroup of A_4. So it is natural to look for an invariant associated with V next.

Let $\zeta_1, \zeta_2, \zeta_3, \zeta_4 \in E$ be the 4 roots of f and set

$$\eta_1 = (\zeta_1 + \zeta_2)(\zeta_3 + \zeta_4)$$
$$\eta_2 = (\zeta_1 + \zeta_3)(\zeta_2 + \zeta_4)$$
$$\eta_3 = (\zeta_1 + \zeta_4)(\zeta_2 + \zeta_3) \ .$$

It is easy to see that

(i) η_1, η_2, and η_3 are all invariant under V;

(ii) no element of S_4 outside V fixes these 3 quantities;

(iii) S_4 permutes η_1, η_2, and η_3.

To verify (ii) you need only check what a 2-cycle, 3-cycle, and 4-cycle do. To verify (iii) it is enough to note that $(1\,2)$ and $(1\,2\,3\,4)$ permute η_1, η_2, and η_3 . Now let

$$r(x) = (x - \eta_1)\,(x - \eta_2)\,(x - \eta_3) \ .$$

Then since the coefficients of r are symmetric in η_1, η_2, and η_3, they are symmetric in $\zeta_1, \zeta_2, \zeta_3, \zeta_4$. Therefore, they must be invariant under G. This means that they lie in F by Corollary 17.3. The polynomial $r \in F[x]$ is called the *cubic resolvent* of f. The splitting field of r is

$$K := F(\eta_1, \eta_2, \eta_3) \ .$$

So $K \subset \mathrm{Fix}\,(V \cap G)$. Since no element of S_4 outside V fixes these three elements, we know that

$$\mathrm{Gal}(E/K) = V \cap G \ ,$$

and therefore

$$K = \mathrm{Fix}\,(V \cap G) \ .$$

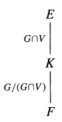

As V is a normal subgroup of S_4, $V \cap G$ is normal in G. So by Theorem 17.7,

$$G/(V \cap G) \cong \mathrm{Gal}(K/F) . \tag{18.2}$$

Also notice that the discriminant of r is the same as the discriminant of f. We have namely,

$$\eta_1 - \eta_2 = \zeta_1\zeta_3 + \zeta_2\zeta_4 - \zeta_1\zeta_2 - \zeta_3\zeta_4 = -(\zeta_1 - \zeta_4)(\zeta_2 - \zeta_3) .$$

Similarly,

$$\eta_2 - \eta_3 = -(\zeta_1 - \zeta_2)(\zeta_3 - \zeta_4)$$
$$\eta_1 - \eta_3 = -(\zeta_1 - \zeta_3)(\zeta_2 - \zeta_4) .$$

Therefore

$$\delta = \prod_{i<j}(\zeta_i - \zeta_j) = -\prod_{i<j}(\eta_i - \eta_j) ,$$

and the discriminants of f and r are the same.

Let us use what we know about cubics to analyze quartics. Suppose first that r is irreducible. Then there are two possibilities: either $\delta \in F$ and $[K : F] = 3$ or $\delta \notin F$ and $[K : F] = 6$. In the first case,

$$3 \mid [E : F] = |G|$$

and $G \subset A_4$. Looking at the list of transitive subgroups of S_4, we see that we must have $G = A_4$. Similarly, in the second case,

$$6 \mid [E : F] = |G|$$

and $G \not\subset A_4$. Therefore $G = S_4$.

Next, suppose that r is reducible. The first possibility is that r splits into linear factors in F, in other words, $\eta_1, \eta_2, \eta_3 \in F$ and $K = F$. By (18.2), this is equivalent to saying that $G \subset V$. The only transitive subgroup of V is V itself. So $G = V$.

The second possibility is that r splits in F into an irreducible quadratic and a linear factor. So $\delta \notin F$ and $[K : F] = 2$. Since $[E : F] \neq 2$, $[E : K] \neq 1$ and

$$|G \cap V| = 2 \text{ or } 4 .$$

In the first case, $[E : F] = 4$ and $G \cong C_4$. In the second case, $[E : F] = 8$ and $\mathrm{Gal}(f) \cong D_4$. Summarizing,

δ	r	$\mathrm{Gal}(r)$	$\mathrm{Gal}(f)$
$\delta \notin F$	r irreducible	S_3	S_4
$\delta \in F$	r irreducible	A_3	A_4
$\delta \in F$	r reducible	$\{1\}$	V
$\delta \notin F$	r reducible	$\mathbb{Z}/2\mathbb{Z}$	D_4 or C_4

In practice you compute the discriminant of f and then check whether it has a square root in F. So we just need a formula for the discriminant of f. Since this is the same as the discriminant of r, we can use our Formula (18.1) for the discriminant of a cubic, if we know what the coefficients of r are. By applying a Tschirnhausen transformation to f (see Exercise 16.10) we can arrange that it has the form

$$f(x) = x^4 + b_2 x^2 + b_1 x + b_0 .$$

Then by Exercise 15.6

$$r(x) = x^3 - 2b_2 x^2 + (b_2^2 - 4b_0)x + b_1^2 .$$

With software to do the calculations for you, it is easy to decide what the Galois group of a quartic is, apart from the ambiguity in the last case in the table.

The Geometry of the Cubic Resolvent

This section requires some understanding of the geometry of conics in $\mathbb{C}^2$.

Suppose we are given a quartic

$$f(x) = x^4 + b_2 x^2 + b_1 x + b_0 ,$$

with $b_2, b_1, b_0 \in \mathbb{C}$. Let us introduce a new variable

$$y := x^2 .$$

Substituting into the quartic gives us a quadratic polynomial in two variables,

$$q_0(x, y) := y^2 + b_2 y + b_1 x + b_0 .$$

If we set

$$q_1(x, y) := x^2 - y ,$$

then solving $f = 0$ is equivalent to solving the pair of equations

$$q_0 = 0$$
$$q_1 = 0$$

simultaneously. We can interpret this geometrically. Each of these equations defines a conic in $\mathbb{C}^2$. They intersect in four points. If $\zeta_1, \zeta_2, \zeta_3, \zeta_4 \in \mathbb{C}$ are the roots of f, then the points of intersection are

$$Q_1 = (\zeta_1, \zeta_1^2), \quad Q_2 = (\zeta_2, \zeta_2^2),$$
$$Q_3 = (\zeta_3, \zeta_3^2), \quad Q_4 = (\zeta_4, \zeta_4^2).$$

The two conics determine a pencil of conics, given by

$$q_t := q_0 + t q_1 = 0, \quad t \in \mathbb{C},$$

which all pass through the four points, Q_1, Q_2, Q_3, Q_4, the basis points of the pencil.

To any quadratic in two variables, one can associate a symmetric 3×3 matrix. If

$$q(x, y) = ax^2 + 2bxy + 2cx + dy^2 + 2ey + f,$$

we can write

$$q(x, y) = (x \ y \ z) \begin{pmatrix} a & b & c \\ b & d & e \\ c & e & f \end{pmatrix} \begin{pmatrix} x \\ y \\ z \end{pmatrix}.$$

So let

$$A_q = \begin{pmatrix} a & b & c \\ b & d & e \\ c & e & f \end{pmatrix}.$$

Then $\det A_q = 0$ if and only if q is degenerate, that is, splits into a product of linear factors. Equivalently, the conic $q = 0$ is a pair of lines.

For what values of t is our pencil of conics degenerate? Well,

$$\det A_{q_t} = \begin{vmatrix} t & 0 & b_1/2 \\ 0 & 1 & (b_2 - t)/2 \\ b_1/2 & (b_2 - t)/2 & b_0 \end{vmatrix}$$

$$= (-t^3 + 2b_2 t^2 + (-b_2^2 + 4b_0)t - b_1^2)/4$$
$$= -r(t)/4.$$

So the pencil degenerates when $t = \eta_i, i = 1, 2, 3$. Each of these degenerate conics is a pair of lines passing through the four points, Q_1, Q_2, Q_3, Q_4. It is not hard to identify exactly which lines belong to each degenerate value.

The line through Q_1 and Q_2 is given by

$$\frac{y - \zeta_1^2}{x - \zeta_1} = \frac{\zeta_2^2 - \zeta_1^2}{\zeta_2 - \zeta_1} = \zeta_1 + \zeta_2,$$

or equivalently, by

$$l_{12}(x, y) := y - (\zeta_1 + \zeta_2) x + \zeta_1 \zeta_2 = 0 .$$

Similarly, the line through Q_3 and Q_4 has equation

$$l_{34}(x, y) := y - (\zeta_3 + \zeta_4) x + \zeta_3 \zeta_4 = 0 .$$

The product of the two linear forms l_{12}, and l_{34} is the degenerate quadratic form

$$\begin{aligned}
l_{12} l_{34}(x, y) &= \big(y - (\zeta_1 + \zeta_2)x + \zeta_1 \zeta_2\big)\big(y - (\zeta_3 + \zeta_4)x + \zeta_3 \zeta_4\big) \\
&= y^2 + (\zeta_1 \zeta_2 + \zeta_3 \zeta_4)y + (\zeta_1 + \zeta_2)(\zeta_3 + \zeta_4)x^2 \\
&\quad - \big(\zeta_1 \zeta_2(\zeta_3 + \zeta_4) + \zeta_3 \zeta_4(\zeta_1 + \zeta_2)\big)x + \zeta_1 \zeta_2 \zeta_3 \zeta_4 \\
&= y^2 + (b_2 - \eta_1)y + \eta_1 x^2 + b_1 x + b_0 \\
&= q_{\eta_1}(x, y) .
\end{aligned}$$

Similarly,

$$l_{13} l_{24} = q_{\eta_2}$$
$$l_{14} l_{23} = q_{\eta_3} .$$

Here is a diagram showing the configuration of the six lines.

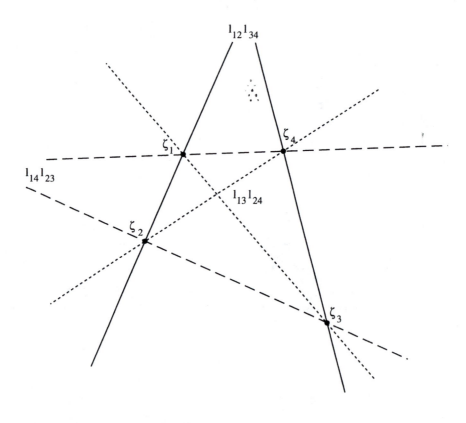

The picture below is a plot of the pencil of real conics for the polynomial $x^4 - 10x^2 + 1$. Its roots are all real. The three degenerate conics are the three pairs of dashed lines. You can also see the four basis points, Q_1, Q_2, Q_3, and Q_4.

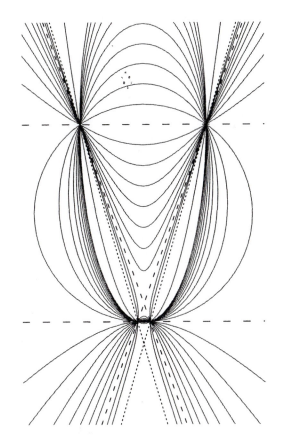

Software

This picture of the pencil of conics was drawn by a package called "Quartics." It also can compute the Galois group of a quartic. First you must load the package:

```
In[1]:= « Quartics.m;
```

A function `CubicResolvent` is defined that computes the cubic resolvent of a quartic, and then applies the criteria in the table above to determine the Galois group (though it cannot distinguish between D_4 and C_4). For example, let

```
In[2]:= f = x^4 - 10x^2 + 1

Out[2]= 1 - 10x^2 + x^4


In[3]:= CubicResolvent[f]

Out[3]= The cubic resolvent is 96x - 20x^2 + x^3.
        Its discriminant is 147456.
        The square root of the discriminant is 384.
        The cubic resolvent factors.
        So the Galois Group is V4.
```

If the quartic has real roots, the function `QuarticPlot` will draw the associated pencil of conics.

```
In[4]:= QuarticPlot[f]
```

Exercises

18.1 What are the possible Galois groups of a reducible cubic?

18.2 What are the possible Galois groups of a reducible quartic?

18.3 Compute the Galois groups of the following rational polynomials:

   (a) $x^4 + x^2 + x + 1$,
   (b) $x^4 + 5x + 5$,
   (c) $x^4 + 8x + 12$.

18.4 Suppose that $g(x) = x^4 + ax^2 + b \in F[x]$ is irreducible. What are the possible Galois groups of g? Give an example of a polynomial g for each.

18.5 Let F be a subfield of $\mathbb{R}$, and h an irreducible quartic in $F[x]$. Suppose that h has exactly two real roots. Prove that $\mathrm{Gal}(h)$ is either S_4 or D_4.

18.6 Let $f(x) = x^4 - x^2 + 4 \in \mathbb{Q}[x]$.

   (a) Show that f splits into the product of two quadratic polynomials over $\mathbb{Q}(\sqrt{5})$. Suggestion: $f(x) = (x^2 + 2)^2 - 5x^2$.
   (b) Find the splitting field E of f.
   (c) What is the Galois group of f?

(d) Verify the Galois correspondence for $\mathrm{Gal}(f)$ and $E/\mathbb{Q}$.

18.7 Let $f(x) = x^4 + ax^3 + bx^2 + ax + 1 \in \mathbb{Q}[x]$ be a reciprocal quartic (see Exercise 15.9). Assume that f is irreducible.

(a) Find the quadratic $g \in \mathbb{Q}[x]$ such that

$$\frac{1}{x^2} f(x) = g\left(x + \frac{1}{x}\right).$$

(b) Let $E/\mathbb{Q}$ be the splitting field of f. Show that g splits in E.

(c) What are the possible Galois groups of f? Give an example of a quartic f for each.

18.8 Let

$$\zeta = \sqrt{2}\sqrt{3}(1 + \sqrt{2})(\sqrt{2} + \sqrt{3}).$$

Verify that ζ is a root of

$$36 - 144x + 108x^2 - 24x^3 + x^4.$$

(You may find the *Mathematica* functions Expand and Simplify useful). Show that $\mathbb{Q}(\sqrt{2}, \sqrt{3})$ is a splitting field for the polynomial over $\mathbb{Q}$. What are its other three roots?

18.9 Let η be the positive square root of ζ. So η satisfies the equation

$$36 - 144x^2 + 108x^4 - 24x^6 + x^8 = 0.$$

Determine the other seven roots and show that the polynomial splits in $\mathbb{Q}(\eta)$. What is its Galois group?

18.10 • Compute the Galois groups of randomly chosen irreducible quartics.

Chapter 19

The General Equation of the nth Degree

Examples

Suppose that $f \in F[x]$ is an irreducible polynomial of degree n with distinct roots in its splitting field. If f is a "generic" polynomial, or one chosen at random, then we would not expect there to be any algebraic relations among its roots, apart from those given by the elementary symmetric polynomials. So the group of symmetries of the roots should be the full permutation group of degree n. In Exercise 18.10, you saw that the Galois group of an irreducible quartic, chosen at random, always seems to be S_4. In fact you probably noticed that it is hard to come up with a quartic whose Galois group is not S_4. If you had a suitable test you would find the same for $n > 4$. In this chapter, we want to give a family of examples of degree p, p prime, with Galois group S_p, and prove that indeed the general polynomial of degree n has Galois group S_n. We shall also give a proof of the fundamental theorem of algebra using Galois theory.

THEOREM 19.1
Let $f \in \mathbb{Q}[x]$ be irreducible of degree p. Suppose that f has precisely 2 nonreal roots. Then $\mathrm{Gal}(f) \cong S_p$.

PROOF   By the fundamental theorem of algebra, f splits in $\mathbb{C}$. So we can assume that the splitting field of f, $E \subset \mathbb{C}$. Let $\zeta_1, \ldots, \zeta_p$ be the roots of f in E. Suppose that ζ_1 and ζ_2 are the two nonreal roots. Since the coefficients of f are real, complex conjugation permutes its roots, and therefore defines an element of $\mathrm{Gal}(f)$ (see Exercise 17.2). It fixes all the real roots and interchanges the two nonreal roots. Thus conjugation corresponds to the transposition $(1\ 2) \in \mathrm{Gal}(f)$.

Now $\mathbb{Q}(\zeta_1) \subset E$, so that

$$[\mathbb{Q}(\zeta_1) : \mathbb{Q}] \,\big|\, [E : \mathbb{Q}] = |\mathrm{Gal}(f)| \ .$$

But

$$[\mathbb{Q}(\zeta_1) : \mathbb{Q}] = \deg f = p \ .$$

Therefore

$$p \mid |\mathrm{Gal}(f)| \, .$$

It follows then from the first Sylow Theorem (see Exercise 11.9) that there is an element of order p in $\mathrm{Gal}(f)$. But by Exercise 6.9, a transposition and an element of order p generate S_p. Therefore $\mathrm{Gal}(f) \cong S_p$. ∎

Examples 19.1

(i) A cubic with $\Delta < 0$ has only one real root (see Exercise 16.12) and therefore has Galois group S_3.

(ii) Let

$$f(x) = x^5 - 6x + 2 \in \mathbb{Q}[x] \, .$$

By the Eisenstein criterion (Theorem 14.4), f is irreducible. And

$$f'(x) = 5x^4 - 6 \, ,$$

which has exactly two real roots. Thus f has at most three real roots. Checking a few values of f shows that it does in fact have three real roots, and therefore two nonreal roots. So

$$\mathrm{Gal}(f) \cong S_5 \, . \quad \square$$

Symmetric Functions

In this section, we are going to prove that in a sense, if there are no nontrivial algebraic relations among the roots of a polynomial, then its Galois group is indeed S_n.

Set $M := F(x_1, \dots, x_n)$, the field of rational functions in the variables $x_1, \dots, x_n$, and let

$$f(x) := (x - x_1) \cdots (x - x_n) \in M[x] \, .$$

Then

$$f(x) = x^n - s_1 x_{n-1} + \cdots + (-1)^n s_n \, ,$$

so that its coefficients actually lie in $F(s_1, \dots, s_n) \subset M$. Clearly there are no algebraic relations among the roots of f except those given by the elementary symmetric polynomials. Furthermore, f is separable and M is the splitting field of f over $F(s_1, \dots, s_n)$. Now the full symmetric group acts on M: for $\alpha \in S_n$ and $g \in M$,

$$(\alpha g)(x_1, \dots, x_n) := g\left(x_{\alpha(1)}, \dots, x_{\alpha(n)}\right) \, .$$

And $L := \mathrm{Fix}(S_n)$ is the field of symmetric functions, which contains $F(s_1, \ldots, s_n)$. So we have the diagram

$$M$$
$$|$$
$$L$$
$$|$$
$$F(s_1, \ldots, s_n)$$

But

$$\mathrm{Gal}(f) = \mathrm{Gal}\big(M/F(s_1, \ldots, s_n)\big) \subset S_n = \mathrm{Gal}(M/L) ,$$

by the fundamental theorem of Galois theory. Therefore

$$\mathrm{Gal}(f) = S_n ,$$

and

$$L = F(s_1, \ldots, s_n) ,$$

in other words, $F(s_1, \ldots, s_n)$ is the field of symmetric functions. As we saw in Corollary 15.1, $s_1, \ldots, s_n$ are algebraically independent. Traditionally an equation like

$$f = 0 ,$$

where the coefficients are algebraically independent, is called a general equation of the nth degree.

REMARK 19.1

(i) As in Exercise 15.11, let

$$\delta(x_1, \ldots, x_n) = \prod_{i<j} (x_i - x_j) .$$

Then $L(\delta) \subset M$ is invariant under A_n, and in fact,

$$\mathrm{Gal}\big(M/L(\delta)\big) = A_n .$$

(ii) Suppose $n = 4$. Set

$$y_1 = (x_1 + x_2)(x_3 + x_4)$$
$$y_2 = (x_1 + x_3)(x_2 + x_4)$$
$$y_3 = (x_1 + x_4)(x_2 + x_3) .$$

Then it is not hard to see that

$$\mathrm{Gal}\big(M/L(y_1, y_2, y_3)\big) = V .$$

So we have the chain of field extensions with the corresponding subgroups of S_4:

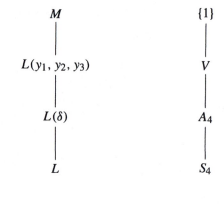

$$M \qquad\qquad \{1\}$$
$$\big|$$
$$L(y_1, y_2, y_3) \qquad\qquad V$$
$$\big|$$
$$L(\delta) \qquad\qquad A_4$$
$$\big|$$
$$L \qquad\qquad S_4$$

∎

The Fundamental Theorem of Algebra

There are many proofs of the fundamental theorem of algebra, some algebraic, some analytic, and some topological. In this section, we shall give a proof that uses Galois theory.

THEOREM 19.2
Any polynomial $f \in \mathbb{C}[x]$ splits in $\mathbb{C}$.

PROOF   First of all, notice that we can assume that f has real coefficients. If not, then consider $f\bar{f} \in \mathbb{R}[x]$. If $\zeta \in \mathbb{C}$ is a root of $f\bar{f}$, then it must be a root of f or of $\bar{f}$. In the latter case, $\bar{\zeta}$ is a root of f. So if $f\bar{f}$ splits in $\mathbb{C}$, then f does as well. Thus we can always replace f by $f\bar{f}$. We can also assume that $\deg f > 1$.

Now let E be the splitting field of f over $\mathbb{R}$. By Exercise 16.9, $[E : \mathbb{R}]$ is even. So $\mathrm{Gal}(f)$ has a nontrivial Sylow 2-subgroup, H_1. Let $K_1 = \mathrm{Fix}(H_1)$. Then $[K_1 : \mathbb{R}]$ is odd. Therefore by Exercise 16.9 again, $K_1 = \mathbb{R}$ and $\mathrm{Gal}(f) = H_1$. In other words, $\mathrm{Gal}(f)$ is a 2-group.

$$E$$
$$H_1 \,\big|$$
$$K_1$$
$$\big|$$
$$\mathbb{R}$$

As we saw in Exercise 11.8, a 2-group has a subgroup of index 2. So let H_2 be of index 2 in $\mathrm{Gal}(f)$. The intermediate field $K_2 = \mathrm{Fix}(H_2)$ is a quadratic extension of $\mathbb{R}$. But according to Exercise 16.1, the only quadratic extension of $\mathbb{R}$ is $\mathbb{C}$. Therefore $K_2 = \mathbb{C}$.

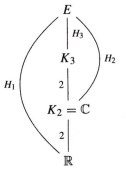

Suppose that $|\mathrm{Gal}(f)| > 2$. Then H_2 is a 2-group, and it too has a subgroup H_3 of index 2. So $K_3 = \mathrm{Fix}(H_3)$ is a quadratic extension of $\mathbb{C}$. But there are no quadratic extensions of $\mathbb{C}$. Therefore $|\mathrm{Gal}(f)| = 2$ and $E = \mathbb{C}$. In other words, f splits in $\mathbb{C}$.

∎

Exercises

19.1 Let

$$h(x) = (x^2 + 12)(x + 4)(x + 2)x(x - 2)(x - 4) - 2 \in \mathbb{Q}[x] \ .$$

(a) Show that h is irreducible. (Suggestion: use Eisenstein's criterion.)

(b) Plot h in the interval $[-5, 5]$.

(c) Prove that h has exactly two nonreal roots and therefore $\mathrm{Gal}(h) \cong S_7$.

19.2 Generalize the construction above to other primes p. You may find it useful to plot examples to get a feel for what is going on.

19.3 Show that in Remark 19.1(i)

$$\mathrm{Gal}\left(M/L(\delta)\right) = A_n \ .$$

19.4 Prove that in Remark 19.1(ii)

$$\mathrm{Gal}\left(M/L(y_1, y_2, y_3)\right) = V \ .$$

19.5 Let G be a group of order n. Show that there exists a subfield $K \subset M = F(x_1, \ldots, x_n)$ such that

$$G \cong \mathrm{Gal}(M/K) \ .$$

19.6 Let F be an infinite field.

(a) Let $n = 2$. Show that for some $a \in F$,

$$M = L(x_1 + ax_2) .$$

Suggestion: First note that there are only finitely many intermediate fields between M and L. Conclude that for some $a, b \in F, a \neq b$,

$$L(x_1 + ax_2) = L(x_1 + bx_2) .$$

(b) Prove that for any n, there exists a rational function $\theta \in M$ such that

$$M = L(\theta) ,$$

in particular, that M/L is a simple extension.

(c) Let E/F be a finite extension. Prove that it is a simple extension.

Chapter 20

Solution by Radicals

In the 16th century, formulas that express the solutions of a cubic equation and a quartic equation in terms of radicals were discovered. For 200 years, mathematicians wondered whether the same could be done for quintics. Around 1800 it was shown that this is impossible. In this chapter, we shall first see how to write the solutions of a cubic in terms of radicals. Then we shall prove that for an equation of degree 5 or greater, no such formulas exist in general. All fields are assumed to be of characteristic 0.

Formulas for a Cubic

There are several ways to derive the formulas for the roots of a cubic. We shall use *Lagrange resolvents* because they will come up again when we discuss equations with cyclic Galois groups.

Suppose F is a field and $f \in F[x]$ an irreducible cubic. According to Exercise 16.10, we can assume that it is of the form

$$f(x) = x^3 + a_1 x + a_0 .$$

If $a_1 = 0$, then the roots are just the cube roots of $-a_0$. So we will also assume that $a_1 \neq 0$. We saw in Chapter 16, page 232, that the splitting field E of f is of degree 3 over $F(\delta)$. We would like to express the roots of f in terms of the cube roots of an element of $F(\delta)$. A Lagrange resolvent does this, at least up to a cube root of unity.

Let

$$\omega = e^{2\pi i/3} = -\frac{1}{2} + \frac{\sqrt{-3}}{2} ,$$

and let ζ_1, ζ_2, and ζ_3 be the roots of f. Then the Lagrange resolvents are

$$\begin{aligned} \xi_1 &:= \zeta_1 + \zeta_2 + \zeta_3 = 0 \\ \xi_2 &:= \zeta_1 + \omega\zeta_2 + \omega^2\zeta_3 \\ \xi_3 &:= \zeta_1 + \omega^2\zeta_2 + \omega\zeta_3 . \end{aligned} \quad (20.1)$$

We compute their cubes in terms of δ:

$$\xi_2^3 = \zeta_1^3 + \zeta_2^3 + \zeta_3^3$$
$$+ 3\omega(\zeta_1^2\zeta_2 + \zeta_2^2\zeta_3 + \zeta_3^2\zeta_1)$$
$$+ 3\omega^2(\zeta_1\zeta_2^2 + \zeta_2\zeta_3^2 + \zeta_3\zeta_1^2)$$
$$+ 6\zeta_1\zeta_2\zeta_3 .$$

By exchanging ω for ω^2, we get a similar expression for ξ_3^3. Now

$$\delta = (\zeta_1 - \zeta_2)(\zeta_2 - \zeta_3)(\zeta_1 - \zeta_3)$$
$$= (\zeta_1^2\zeta_2 + \zeta_2^2\zeta_3 + \zeta_3^2\zeta_1) - (\zeta_1\zeta_2^2 + \zeta_2\zeta_3^2 + \zeta_3\zeta_1^2) .$$

Since we can write

$$\xi_2^3 = \zeta_1^3 + \zeta_2^3 + \zeta_3^3$$
$$+ \frac{3}{2}(\omega + \omega^2)\left[(\zeta_1^2\zeta_2 + \zeta_2^2\zeta_3 + \zeta_3^2\zeta_1) + (\zeta_1\zeta_2^2 + \zeta_2\zeta_3^2\zeta_3\zeta_1^2)\right]$$
$$+ \frac{3}{2}(\omega - \omega^2)\left[(\zeta_1^2\zeta_2 + \zeta_2^2\zeta_3 + \zeta_3^2\zeta_1) - (\zeta_1\zeta_2^2 + \zeta_2\zeta_3^2\zeta_3\zeta_1^2)\right] + 6\zeta_1\zeta_2\zeta_3$$

and

$$\omega + \omega^2 = -1 \quad , \quad \omega - \omega^2 = \sqrt{-3} ,$$

we have

$$\xi_2^3 = \sum_j \zeta_j^3 - \frac{3}{2}\sum_{j\neq k} \zeta_j^2\zeta_k + 6\zeta_1\zeta_2\zeta_3 + \frac{3}{2}\sqrt{-3}\,\delta . \qquad (20.2)$$

The first three terms above are symmetric in ζ_1, ζ_2, and ζ_3. We can therefore express them in terms of the elementary symmetric polynomials, whose values are $-a_2 = 0$, a_1, and $-a_0$:

$$\sum_j \zeta_j^3 = -a_2^3 + 3a_1a_2 - 3a_0 = -3a_0$$

$$\sum_{j\neq k} \zeta_j^2\zeta_k = -a_1a_2 + 3a_0 \qquad = 3a_0 .$$

Substituting into (20.2) gives

$$\xi_2^3 = -\frac{27}{2}a_0 + \frac{3}{2}\sqrt{-3}\,\delta .$$

Similarly,

$$\xi_3^3 = -\frac{27}{2}a_0 - \frac{3}{2}\sqrt{-3}\,\delta .$$

So we can take cube roots of the right-hand sides of these two equations to obtain ξ_2 and ξ_3.

$$\xi_2 = \sqrt[3]{-\frac{27}{2}a_0 + \frac{3}{2}\sqrt{-3}\,\delta} \tag{20.3}$$

$$\xi_3 = \sqrt[3]{-\frac{27}{2}a_0 - \frac{3}{2}\sqrt{-3}\,\delta}\,.$$

There are three choices for each cube root. But we must keep in mind that

$$\xi_2\xi_3 = (\zeta_1^2 + \zeta_2^2 + \zeta_3^2) + (\omega + \omega^2)(\zeta_1\zeta_2 + \zeta_2\zeta_3 + \zeta_1\zeta_3)$$
$$= (\zeta_1^2 + \zeta_2^2 + \zeta_3^2) - (\zeta_1\zeta_2 + \zeta_2\zeta_3 + \zeta_1\zeta_3)$$
$$= -3a_1\,.$$

So each value of ξ_2 determines a value of ξ_3. Lastly, we can use Equations (20.1) to express the roots of f in terms of the Lagrange resolvents:

$$\zeta_1 = \frac{1}{3}(\xi_2 + \xi_3)$$

$$\zeta_2 = \frac{1}{3}(\omega^2\xi_2 + \omega\xi_3) \tag{20.4}$$

$$\zeta_3 = \frac{1}{3}(\omega\xi_2 + \omega^2\xi_3)\,.$$

Choosing a different cube root in (20.3) means multiplying ξ_2 by ω (or ω^2) and ξ_3 by ω^2 (or ω). This does not change $\{\zeta_1, \zeta_2, \zeta_3\}$. These are *Cardano's formulas* for the roots of a cubic. Notice that if $\omega \in F$, then

$$E = F(\delta, \xi_2) = F(\delta, \xi_3)\,.$$

As was pointed out in Exercise 17.7,

$$\mathrm{Gal}\left(E/F(\delta)\right) \cong A_3\,,$$

which is cyclic of order 3.

Example 20.1

Take $f(x) = x^3 + x - 2$. Then by Exercise 16.11, $\delta = 4\sqrt{-7}$, which gives

$$\xi_2^3 = 27 + 6\sqrt{21} \quad \text{and} \quad \xi_3^3 = 27 - 6\sqrt{21}\,.$$

Taking the real cube roots, we have

$$\xi_2 = \sqrt[3]{27 + 6\sqrt{21}} \quad \text{and} \quad \xi_3 = \sqrt[3]{27 - 6\sqrt{21}}\,.$$

Formulas (20.4) then give us the roots of f:

$$\sqrt[3]{27 + 6\sqrt{21}} + \sqrt[3]{27 - 6\sqrt{21}},$$

$$\omega^2\sqrt[3]{27 + 6\sqrt{21}} + \omega\sqrt[3]{27 - 6\sqrt{21}},$$

$$\omega\sqrt[3]{27 + 6\sqrt{21}} + \omega^2\sqrt[3]{27 - 6\sqrt{21}}.$$

But obviously 1 is a root of f, and since

$$x^3 + x - 2 = (x - 1)(x^2 + x + 2),$$

the other two roots are just

$$-\frac{1}{2} \pm \frac{\sqrt{-7}}{2}.$$

This makes it clear that Cardano's formulas can easily produce expressions for the roots which are rather obscure. ☐

Cyclic Extensions

In Example 17.2, we calculated the Galois group of $x^p - 2$ over $\mathbb{Q}$. The result is that after adjoining the pth roots of 1 to $\mathbb{Q}$, the Galois group is cyclic of order p. This is typical.

THEOREM 20.1
Let F be a field of characteristic 0 and let $f(x) = x^n - a \in F[x]$. Suppose that F contains the nth roots of 1. Then $\mathrm{Gal}(f)$ is cyclic and its order divides n.

PROOF   Let $E = F[x]/(x^n - a)$. If $\zeta \in E$ is a root of f, then so is $\omega\zeta$, if $\omega \in F$ is an nth root of 1. Thus E is the splitting field of f over F.

If $\alpha \in \mathrm{Gal}(E/F)$, then α is determined by $\alpha(\zeta)$. Furthermore, $\alpha(\zeta)$ must be of the form $\omega\zeta$ for some nth root of 1, ω. Therefore let us define $\alpha_\omega \in \mathrm{Gal}(f)$ by

$$\alpha_\omega(\zeta) = \omega\zeta,$$

where ω is an nth root of 1. This defines a mapping

$$\mu_n \to \mathrm{Gal}(f),$$

(see Example 6.3(iii)), which we already know is surjective. It is a homomorphism. Suppose that ω_1 and ω_2 are nth roots of 1. Then

$$\alpha_{\omega_1\omega_2}(\zeta) = (\omega_1\omega_2)\zeta = \alpha_{\omega_1}\left(\alpha_{\omega_2}(\zeta)\right).$$

So
$$\alpha_{\omega_1\omega_2} = \alpha_{\omega_1}\alpha_{\omega_2} .$$

Therefore $\text{Gal}(f)$ is a quotient of a cyclic group of order n. So it too is cyclic, and its order divides n. ∎

An extension E/F, where E is the splitting field of a separable polynomial in $F[x]$, is called *cyclic* if $\text{Gal}(E/F)$ is cyclic. If E/F is the splitting field of an irreducible cubic as in the previous section, then $E/F(\delta)$ is a cyclic extension. We showed there that $E = F(\delta, \xi_2)$ with $\xi_2^3 \in F(\delta)$. In general if E/F is cyclic, then $E = F(\sqrt[n]{a})$ for some $a \in F$. This can be proved as above by finding a Lagrange resolvent ξ with $\xi^n \in F$ and $E = F(\xi)$.

DEFINITION 20.1   *Suppose that E/F is a cyclic extension, and α is a generator of its Galois group. Assume that F contains n distinct nth roots of 1. Then for $\zeta \in E$ and $\omega \in F$, an nth root of 1, we define the Lagrange resolvent*

$$(\omega, \zeta) := \zeta + \omega\alpha(\zeta) + \cdots + \omega^{n-1}\alpha^{n-1}(\zeta) . \tag{20.5}$$

Equations (20.1) give Lagrange resolvents for $n = 3$.

THEOREM 20.2
Let F be a field containing the nth roots of 1. Suppose that E/F is a cyclic extension. Then
$$E = F\left(\sqrt[n]{a}\right) ,$$
for some $a \in F$.

PROOF   Let ω be a primitive nth root of 1. First notice that

$$\alpha\big((\omega, \zeta)\big) = \alpha(\zeta) + \omega\alpha^2(\zeta) + \cdots + \omega^{n-1}\zeta$$
$$= \omega^{-1}(\omega, \zeta) .$$

This implies that if $(\omega, \zeta) \neq 0$ and $1 \leq i < n$, then

$$\alpha\big((\omega, \zeta)^i\big) = \omega^{-i}(\omega, \zeta)^i \neq (\omega, \zeta)^i ,$$

and therefore

$$(\omega, \zeta)^i \notin F .$$

However

$$\alpha\big((\omega, \zeta)^n\big) = (\omega, \zeta)^n ,$$

so that

$$(\omega, \zeta)^n \in F .$$

Thus if $(\omega, \zeta) \neq 0$,

$$E = F\big((\omega, \zeta)\big),$$

and we can take

$$a = (\omega, \zeta)^n.$$

It remains to show that ζ can be found with $(\omega, \zeta) \neq 0$.

Now for each i, $0 \leq i < n$, we can replace ω by ω^i in (20.5). This gives us a system of n equations:

$$
\begin{pmatrix} (1, \zeta) \\ (\omega, \zeta) \\ \vdots \\ (\omega^{n-1}, \zeta) \end{pmatrix}
=
\begin{pmatrix} 1 & 1 & \cdots & 1 \\ 1 & \omega & \cdots & \omega^{n-1} \\ \vdots & \vdots & & \vdots \\ 1 & \omega^{n-1} & \cdots & (\omega^{n-1})^{n-1} \end{pmatrix}
\begin{pmatrix} \zeta \\ \alpha(\zeta) \\ \vdots \\ \alpha^{n-1}(\zeta) \end{pmatrix}
$$

It is easy to see that

$$
\begin{vmatrix} 1 & 1 & \cdots & 1 \\ 1 & \omega & \cdots & \omega^{n-1} \\ \vdots & \vdots & & \vdots \\ 1 & \omega^{n-1} & \cdots & (\omega^{n-1})^{n-1} \end{vmatrix}
= \prod_{j>k} \left(\omega^j - \omega^k \right) \neq 0.
$$

Therefore the matrix is invertible and we can write

$$
\begin{pmatrix} \zeta \\ \alpha(\zeta) \\ \vdots \\ \alpha^{n-1}(\zeta) \end{pmatrix}
=
\begin{pmatrix} 1 & 1 & \cdots & 1 \\ 1 & \omega & \cdots & \omega^{n-1} \\ \vdots & \vdots & & \vdots \\ 1 & \omega^{n-1} & \cdots & (\omega^{n-1})^{n-1} \end{pmatrix}^{-1}
\begin{pmatrix} (1, \zeta) \\ (\omega, \zeta) \\ \vdots \\ (\omega^{n-1}, \zeta) \end{pmatrix}.
$$

It follows that if $\zeta \neq 0$, then for some i, $(\omega^i, \zeta) \neq 0$. Replacing ω by ω^i if necessary, we then have what we want. ∎

Solution by Radicals in Higher Degrees

Formulas (20.3) and (20.4) express the solutions of the cubic in terms of the square roots of $\Delta \in F$ and the cube roots of $-(27/2)a_0 + (3/2)\sqrt{-3}\sqrt{\Delta} \in F(\sqrt{\Delta})$. Can one find similar formulas for equations of higher degree? In other words, can one build up an expression for the solutions by starting with an mth root of an element $a \in F$, then taking an nth root of an element $b \in F(\sqrt[m]{a})$, and so on? In terms of the splitting field of the equation, this means that it should be built up as a sequence of "radical" extensions:

$$E$$

$$F(\sqrt[m]{a})(\sqrt[n]{b})$$

$$F(\sqrt[m]{a})$$

$$F$$

Example 20.2
Let $f(x) = x^6 - 4x^3 + 1 \in \mathbb{Q}[x]$. First note that $g(y) = y^2 - 4y + 1$ has roots
$2 \pm \sqrt{3}$. Therefore, the roots of f are the cube roots of $2 + \sqrt{3}$ and $2 - \sqrt{3}$. Now

$$2 - \sqrt{3} = \frac{1}{2 + \sqrt{3}} \qquad \text{and} \qquad \sqrt[3]{2 - \sqrt{3}} = \frac{1}{\sqrt[3]{2 + \sqrt{3}}}$$

(cf. Exercise 15.9). Therefore the splitting field of f is $\mathbb{Q}(\sqrt[3]{2 + \sqrt{3}}, \omega)$, where ω
is a primitive cube root of 1. This can be built up as

$$\mathbb{Q}\left(\sqrt[3]{2 + \sqrt{3}}, \omega\right)$$

$$\mathbb{Q}\left(\sqrt[3]{2 + \sqrt{3}}\right)$$

$$\mathbb{Q}(\sqrt{3})$$

$$\mathbb{Q}$$

□

As mentioned at the beginning of this chapter, this can be done in general for
quartics, but not for equations of degree greater than 4. More precisely, we shall
prove that for the general equation of degree greater than 4, that is, one whose Galois

group is S_n, no such formulas can exist. In fact, it can be shown that if such formulas do exist for a specific equation of degree p, where $p \geq 5$ is prime, then its Galois group must be a subgroup of the Frobenius group $F_{(p-1)p}$. Since the Frobenius group is a rather small subgroup of S_p, this result suggests that the general equation of say, degree 5, is very far from being solvable by radicals.

First we need to know that we can assume that the base field F contains all the roots of unity that we need.

LEMMA 20.1

Suppose $f \in F[x]$ and let E be the splitting field of f. Let ω be a primitive mth root of 1. Then

$$\mathrm{Gal}\left(E(\omega)/F(\omega)\right) \cong \mathrm{Gal}\left(E/E \cap F(\omega)\right)$$

and

$$\mathrm{Gal}\left(E/E \cap F(\omega)\right) \lhd \mathrm{Gal}(E/F) .$$

PROOF   Here is a diagram of the field extensions:

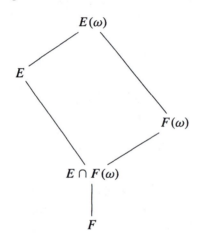

Since E/F is the splitting field of f, we have a surjective homomorphism

$$\mathrm{res} : \mathrm{Gal}\left(E(\omega)/F\right) \to \mathrm{Gal}(E/F)$$

by Theorem 17.7. It is defined by restricting an element of $\mathrm{Gal}\left(E(\omega)/F\right)$ to E, and thus its kernel consists of automorphisms whose restriction to E is trivial.

Now $\mathrm{Gal}\left(E(\omega)/F(\omega)\right) \subset \mathrm{Gal}\left(E(\omega)/F\right)$ and the composed mapping

$$\mathrm{Gal}\left(E(\omega)/F(\omega)\right) \hookrightarrow \mathrm{Gal}\left(E(\omega)/F\right) \to \mathrm{Gal}(E/F)$$

is injective: an automorphism of $E(\omega)$ that is trivial on $F(\omega)$ and on E, is trivial. Furthermore, the image H lies in $\mathrm{Gal}\left(E/E \cap F(\omega)\right)$. We want to see that it is all of $\mathrm{Gal}\left(E/E \cap F(\omega)\right)$. Suppose $\zeta \in E$ is fixed by H. If we regard ζ as an element of

$E(\omega)$, this means that it is fixed by $\mathrm{Gal}\left(E(\omega)/F(\omega)\right)$, and therefore lies in $F(\omega)$. So $\zeta \in E \cap F(\omega)$ and

$$\mathrm{Fix}\ H = E \cap F(\omega)\ .$$

It follows that

$$H = \mathrm{Gal}\left(E/E \cap F(\omega)\right)\ .$$

To see that it is a normal subgroup of $\mathrm{Gal}(E/F)$, take $\beta \in \mathrm{Gal}\left(E/E \cap F(\omega)\right)$ and $\alpha \in \mathrm{Gal}(E/F)$. Choose $\beta' \in \mathrm{Gal}\left(E(\omega)/F(\omega)\right)$ and $\alpha' \in \mathrm{Gal}(E(\omega)/F)$ such that

$$\mathrm{res}(\alpha') = \alpha \quad \text{and} \quad \mathrm{res}(\beta') = \beta\ .$$

Then

$$\alpha'\beta'{\alpha'}^{-1} \in \mathrm{Gal}\left(E(\omega)/F(\omega)\right)$$

since $\mathrm{Gal}\left(E(\omega)/F(\omega)\right) \triangleleft \mathrm{Gal}(E(\omega)/F)$ by Theorem 17.7. Therefore

$$\alpha\beta\alpha^{-1} = \mathrm{res}(\alpha'\beta'{\alpha'}^{-1}) \in \mathrm{Gal}\left(E/E \cap F(\omega)\right)\ ,$$

and

$$\mathrm{Gal}\left(E/E \cap F(\omega)\right) \triangleleft \mathrm{Gal}(E/F)\ . \quad \blacksquare$$

COROLLARY 20.1
If $E \not\subset F(\omega)$ and if $\mathrm{Gal}(E/F)$ is a simple group, then

$$\mathrm{Gal}\left(E(\omega)/F(\omega)\right) \cong \mathrm{Gal}(E/F)\ .$$

We can now prove the main theorem. The heart of the proof is the fact that A_n is a simple group if $n \geq 5$.

THEOREM 20.3
Let $f \in F[x]$ be an irreducible polynomial of degree $n \geq 5$ with Galois group S_n. Then f cannot be solved by taking radicals.

PROOF Let E be the splitting field of f over F. Suppose that f can be solved by radicals. First we adjoin δ, the square root of the discriminant, to F. By Theorem 17.8 we know that

$$\mathrm{Gal}\left(E/F(\delta)\right) = A_n\ .$$

So we have the diagram:

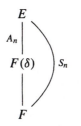

Suppose that the next step is to adjoin $\sqrt[m]{a}$ to $F(\delta)$. Thus $F(\delta, \sqrt[m]{a})$ is a subfield of E. We can assume that F contains m distinct mth roots of 1. Otherwise we adjoin ω, a primitive mth root of 1 and replace F and E by $F(\omega)$ and $E(\omega)$, respectively. Then by the lemma above, $\mathrm{Gal}\left(E(\omega)/F(\delta, \omega)\right) \cong A_n$. Set

$$H = \mathrm{Gal}\left(E/F(\delta, \sqrt[m]{a})\right) .$$

Since $F(\sqrt[m]{a}, \delta)$ is the splitting field of $x^m - a$ over $F(\delta)$, by Theorem 17.7, H is a normal subgroup of A_n, and

$$\mathrm{Gal}\left(F(\delta, \sqrt[m]{a})/F(\delta)\right) \cong A_n/H .$$

Here is the diagram:

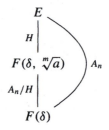

But as we saw in Theorem 12.4, A_n is simple for $n > 4$. This means that either $H = A_n$ and $F(\delta, \sqrt[m]{a})/F(\delta)$ is trivial, or H is trivial and $A_n = \mathrm{Gal}\left(F(\delta, \sqrt[m]{a})/F(\delta)\right)$, which is cyclic. Both are impossible. Therefore f cannot be solved by taking radicals.

■

COROLLARY 20.2 (Corollary to Proof)
   If $\mathrm{Gal}(f) \cong A_n$, $n \geq 5$, *then* f *is not solvable by radicals.*

One can go further and algebraically characterize the Galois groups of polynomials that can be solved by radicals. They are a class of groups called *solvable groups*.

DEFINITION 20.2   *A group* G *is solvable if it has a composition series*

$$G = H_0 \rhd H_1 \rhd \cdots \rhd H_{n-1} \rhd H_n = \{1\} ,$$

where H_i/H_{i+1} *is a cyclic group of prime order, for* $0 \leq i < n$.

These turn out to be of great significance in group theory as well.

Calculations

Mathematica has a built-in function Solve for solving equations (algebraic or otherwise). It manages to avoid some of the foibles of Cardano's formulas.

Examples 20.3

(i) $In[1] := Solve[x^3 - 3x + 5 == 0, x]$

$$Out[1] = \left\{\left\{ x \rightarrow -\left(\frac{2}{5 - \sqrt{21}}\right)^{1/3} - \left(\frac{1}{2}(5 - \sqrt{21})\right)^{1/3} \right\}, \right.$$
$$\left\{ x \rightarrow \frac{1}{2}(1 + i\sqrt{3})\left(\frac{1}{2}(5 - \sqrt{21})\right)^{1/3} \right.$$
$$\left. + \frac{1 - i\sqrt{3}}{2^{2/3}(5 - \sqrt{21})^{1/3}} \right\},$$
$$\left\{ x \rightarrow \frac{1}{2}(1 - i\sqrt{3})\left(\frac{1}{2}(5 - \sqrt{21})\right)^{1/3} \right.$$
$$\left.\left. + \frac{1 + i\sqrt{3}}{2^{2/3}(5 - \sqrt{21})^{1/3}} \right\}\right\}$$

(ii) $In[2] := Solve[x^4 - 3x + 3 == 0, x]$

$$Out[2] = \left\{\left\{ x \rightarrow \frac{i\sqrt{3}}{2} - \frac{1}{2}\sqrt{3 - 2i\sqrt{3}} \right\}, \right.$$
$$\left\{ x \rightarrow \frac{i\sqrt{3}}{2} + \frac{1}{2}\sqrt{3 - 2i\sqrt{3}} \right\},$$
$$\left\{ x \rightarrow -\frac{i\sqrt{3}}{2} - \frac{1}{2}\sqrt{3 - 2i\sqrt{3}} \right\},$$
$$\left.\left\{ x \rightarrow -\frac{i\sqrt{3}}{2} + \frac{1}{2}\sqrt{3 - 2i\sqrt{3}} \right\}\right\}$$ ▯

Exercises

20.1 Find the roots of $x^3 - 3x + 3$.

20.2 Let ω be a primitive nth root of 1. Show that

$$\begin{vmatrix} 1 & 1 & \cdots & 1 \\ 1 & \omega & \cdots & \omega^{n-1} \\ \vdots & \vdots & & \vdots \\ 1 & \omega^{n-1} & \cdots & (\omega^{n-1})^{n-1} \end{vmatrix} = \prod_{n \geq k > j \geq 1} \left(\omega^k - \omega^j\right).$$

20.3 Suppose that E is an extension field of F and that E does not contain any mth roots of 1. Let ω be a primitive mth root of 1 in some extension of F. Prove that

$$[F(\omega) : F] = [E(\omega) : E] .$$

20.4 Find the minimal polynomial of $\cos(2\pi/11)$. Show that its Galois group is cyclic of degree 5. Prove that it can be solved by radicals.

20.5 Let $f(x) = x^8 - 2x^4 - 2 \in \mathbb{Q}[x]$. Determine the splitting field E of f. Prove that f can be solved by radicals by showing how E can be built up as a sequence of radical extensions. What is the corresponding sequence of subgroups of $\mathrm{Gal}(f)$? Demonstrate that $\mathrm{Gal}(f)$ is a solvable group.

20.6 A *normal* series for a group G is a sequence of subgroups

$$G = H_0 \rhd H_1 \rhd \cdots \rhd H_{n-1} \rhd H_n = \{1\} .$$

Prove that a finite group G is solvable if and only it has a normal series where all the factors $H_j/H_{j+1}, 0 \leq j < n$, are cyclic. Prove that this also holds if and only if it has a normal series where all the factors H_j/H_{j+1} are abelian.

Chapter 21

Ruler-and-Compass Constructions

Introduction

In this chapter, we are going to look at geometric constructions you can make with a ruler and compass. To be more precise, we should say straight edge and compass because we are only using the ruler to draw straight lines, not to make measurements. It is not hard to bisect a line segment or an angle. You can also construct an equilateral triangle or a square. But can you trisect an angle? What other regular polygons can you construct?

Such geometric constructions can be translated into algebraic problems involving the solution of polynomial equations. As an example, let us look at the question of trisecting an angle. Suppose we want to trisect an angle τ, $0 < \tau < \pi/2$. We put the vertex of the angle at the point $(0, 0)$ and one arm along the x-axis. The other arm then passes through the point $(\cos \tau, \sin \tau)$. Since

$$\sin \tau = \sqrt{1 - \cos^2 \tau} \,,$$

the angle is determined by the real number $\cos \tau$. Similarly, the angle $\tau/3$ can be measured from the x-axis and then it is determined by $\cos(\tau/3)$.

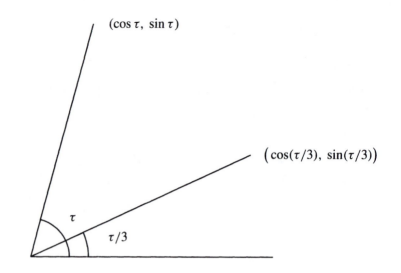

Now we have the well-known triple angle formula:

$$\cos \tau = \cos 3(\tau/3) = 4 \cos^3(\tau/3) - 3 \cos(\tau/3) .$$

In other words, $\cos(\tau/3)$ is a solution of the real cubic

$$4x^3 - 3x - \cos \tau = 0 .$$

Trisecting the given angle τ is equivalent to solving this equation.

We are going to see that with ruler and compass we can perform constructions that solve sequences of quadratic equations, but not cubic equations. For this reason, trisecting an angle with such constructions is in general impossible. Of course there are special angles, like $\pi/2$, which you can trisect in this way.

The fundamental criterion for constructibility in the next section only uses the results on field extensions from Chapter 16. It does not need any Galois theory. The discussion of periods and explicit equations arising in the construction of regular polygons in the following section does use Galois theory.

Algebraic Interpretation

We imagine that to start, we are given some points in the plane and some lengths. Given a point, we can construct its projections on the x- and y-axes, in other words, its coordinates. Conversely, we can construct a point from its x- and y-coordinates. So we can assume that our given data simply consists of some real numbers $a_1, \dots, a_r$. For the same reasons, when we ask what can be constructed from this data, we need only look at the lengths that can be constructed. We assume that we have chosen a unit length. Then let $L \subset \mathbb{R}$ be the set of all lengths that can be constructed with ruler and compass.

THEOREM 21.1
L is a field containing $\mathbb{Q}(a_1, \ldots, a_r)$. *If* $b \in L$, *then* $\sqrt{b} \in L$.

PROOF   It is clear how to add two given lengths using a compass, or how to subtract one from the other. How can one multiply two lengths b_1 and b_2? Mark the unit length on the x-axis, and the length b_1 on the y-axis. Join the two points.

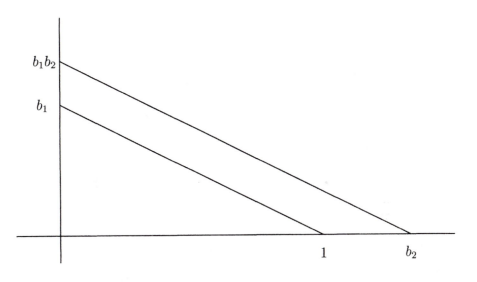

Now mark the length b_2 on the x-axis. Draw a line through this point parallel to the first line. It will meet the y-axis at the point $b_1 b_2$. To construct b_1/b_2, switch the roles of 1 and b_2. Therefore L is a field. Since it contains $a_1, \ldots, a_r$ and contains $\mathbb{Q}$, it contains $\mathbb{Q}(a_1, \ldots, a_r)$.

To construct the square root of a given length b, mark 1 and $1 + b$ on a line. Draw a semicircle with diameter $1 + b$. Draw the perpendicular from the point 1 to the semicircle, and let its length be c.

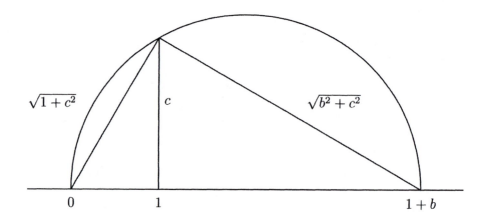

The hypotenuses of the two right-angled triangles on the diameter have lengths $\sqrt{1 + c^2}$ and $\sqrt{b^2 + c^2}$ as shown. But the large triangle is also right-angled. Therefore

$$(1 + c^2) + (b^2 + c^2) = (1 + b)^2 ,$$

which implies that

$$c = \sqrt{b} . \quad \blacksquare$$

For example, in Exercise 16.4 you saw that the minimal polynomial of $\cos(2\pi/5)$ is

$$4x^2 + 2x - 1 .$$

Therefore

$$\cos(2\pi/5) = \frac{\sqrt{5} - 1}{4} ,$$

and we can construct $\cos(2\pi/5)$ with ruler and compass. This is the essential step in constructing a regular pentagon. In [6, Chapter 2], there is a more direct construction.

The theorem above can be interpreted in the following way. Suppose $F \subset L$ is a subfield and $b \in F$. Then any length in $F(\sqrt{b})$ is constructible. So if a real number ζ lies in an extension field of $\mathbb{Q}(a_1, \ldots, a_r)$, which can be built up as a sequence of quadratic extensions, then ζ is constructible. On the other hand, suppose that we have a construction that starts with $a_1, \ldots, a_r$ and gives us a length $\zeta \in L$. It is built up out of three basic steps:

(i) intersect two lines;

(ii) intersect a line with a circle;

(iii) intersect two circles.

From an algebraic point of view, we are beginning with the base field $\mathbb{Q}(a_1, \ldots, a_r)$ and each step gives us a field extension. Let us analyze what sort of extension we get in each case.

Let F be a subfield of $\mathbb{R}$ containing $a_1, \ldots, a_r$. Suppose we have a line passing through two points whose coordinates lie in F. Then its equation will be of the form

$$ax + by + c = 0 ,$$

where $a, b, c \in F$. If we have a second such line,

$$dx + ey + f = 0 ,$$

then the solution (x, y) of the pair of equations is the point of intersection. And x and y belong to F. So when we intersect two lines we make no field extension.

Next suppose we are given a circle whose radius is in F, and whose centre is a point with coordinates in F. Then it has an equation of the form

$$x^2 + y^2 + dx + ey + f = 0 ,$$

where $d, e, f \in F$. If it meets a line

$$ax + by + c = 0 ,$$

$a, b, c \in F$, then to find the coordinates of the two points of intersection, we solve the linear equation, for x, say, (assuming that $a \neq 0$), and substitute into the quadratic. This gives us a quadratic equation in one variable y. If we adjoin a square root of its discriminant Δ to F, then the two solutions for y, and the corresponding values of x lie in $F(\sqrt{\Delta})$. So constructing the points of intersection of a circle and a line corresponds to making a quadratic extension.

Lastly, suppose we are given two circles, determined by data in F, which meet:

$$x^2 + y^2 + ax + by + c = 0$$
$$x^2 + y^2 + dx + ey + f = 0 ,$$

$a, b, c, d, e, f \in F$. Subtracting, we get the equation

$$(a - d)x + (b - e)y + (c - f) = 0 .$$

A common solution of the two quadratic equations is a solution of the linear equation. Therefore the linear equation gives the line passing through the points of intersection of the two circles. So finding the points of intersection of two circles can be reduced to finding the intersection of a circle and a line. Therefore, the coordinates of the points of intersection will lie in a quadratic extension of F too.

Summarizing, we have the following result:

THEOREM 21.2

Given $a_1, \ldots, a_r \in \mathbb{R}$, we can construct a real number ζ from $a_1, \ldots, a_r$ using ruler and compass if and only if ζ lies in an extension $E/\mathbb{Q}(a_1, \ldots, a_r)$, which can be built up as a sequence of quadratic extensions.

COROLLARY 21.1

If ζ is constructible, then $[\mathbb{Q}(a_1, \ldots, a_r, \zeta) : \mathbb{Q}(a_1, \ldots, a_r)]$ is a power of 2.

PROOF   If ζ is constructible, then $\zeta \in E$, where E is an extension of $\mathbb{Q}(a_1, \ldots, a_r)$, which can be built up as a sequence of quadratic extensions. Thus

$$[\mathbb{Q}(a_1, \ldots, a_r, \zeta) : \mathbb{Q}(a_1, \ldots, a_r)] \mid [E : \mathbb{Q}(a_1, \ldots, a_r)] = 2^k \,,$$

for some $k \geq 0$.   ∎

Examples 21.1

(i) Suppose that we want to trisect the angle $\pi/3$. As discussed in the Introduction, we begin with $\cos(\pi/3) = 1/2 \in \mathbb{Q}$. We want to construct $\cos(\pi/9)$, which is a root of

$$4x^3 - 3x - 1/2 = 0 \,.$$

It is not hard to see that this polynomial is irreducible over $\mathbb{Q}$. Therefore

$$[\mathbb{Q}(\cos(\pi/9)) : \mathbb{Q}] = 3 \,.$$

The theorem then tells us that $\cos(\pi/9)$ is not constructible.

(ii) Suppose we want to duplicate the cube. This means that we begin with a cube with unit volume, and want to construct one with volume 2. So the sides of the new cube must have length $\sqrt[3]{2}$, and this is the number we want to construct. But

$$[\mathbb{Q}(\sqrt[3]{2}) : \mathbb{Q}] = 3 \,.$$

Therefore $\sqrt[3]{2}$ is not constructible, and it is not possible to duplicate the cube with a ruler-and-compass construction.   ☐

Construction of Regular Polygons

If we inscribe a regular n-gon P_n in the unit circle about the origin in $\mathbb{R}^2$, with one vertex at the point $(1, 0)$, then the other vertices lie at the points

$$\left\{ \left(\cos(2\pi k/n), \sin(2\pi k/n) \right) \mid 0 < k < n \right\} \,.$$

If we can construct $\left(\cos(2\pi/n), \sin(2\pi/n) \right)$, then we can construct the other vertices from it. This is precisely the case when its components are constructible. Since

$$\sin(2\pi/n) = \sqrt{1 - \cos^2(2\pi/n)} \,,$$

we see that $\sin(2\pi/n)$ is constructible if and only if $\cos(2\pi/n)$ is. In Exercise 17.10 you saw that

$$[\mathbb{Q}(\cos(2\pi/n)) : \mathbb{Q}] = \varphi(n)/2 \, .$$

Therefore by Corollary 21.1, if P_n is constructible, $\varphi(n)$ must be a power of 2.

Conversely, suppose that $\varphi(n)$ is a power of 2. Since

$$\varphi(n) = [E_n : \mathbb{Q}] = |\text{Gal}\,(\Phi_n)| \, ,$$

and

$$\text{Gal}\,(\Phi_n) \cong (\mathbb{Z}/n\mathbb{Z})^\times$$

this tells us that $\text{Gal}(\Phi_n)$ is an abelian 2-group (see page 250). Therefore $\text{Gal}\big(\mathbb{Q}(\cos(2\pi/n))/\mathbb{Q}\big)$ is as well. Then it follows from the first Sylow Theorem (see Exercise 11.8) or from the classification of finite abelian groups that there is a chain of subgroups

$$\{1\} =: G_0 < G_1 < \cdots < G_r := \text{Gal}\big(\mathbb{Q}(\cos(2\pi/n))/\mathbb{Q}\big) \, ,$$

where G_j/G_{j-1} has order 2, for all $j \geq 1$. Corresponding to this is a tower of quadratic extensions

$$Q(\cos(2\pi/n)) \supset \text{Fix}(G_1) \supset \cdots \supset \mathbb{Q} \, .$$

Therefore by our criterion for constructibility, $\cos(2\pi/n)$ is constructible.

So when is $\varphi(n)$ a power of 2? In Exercise 5.24, we found that if

$$n = p_1^{j_1} \cdots p_r^{j_r} \, ,$$

with $p_1, \ldots, p_r$ distinct primes, then

$$\varphi(n) = p_1^{j_1-1} (p_1 - 1) \cdots p_r^{j_r-1} (p_r - 1) \, .$$

So $\varphi(n)$ is a power of 2 if and only if n factors as

$$n = 2^j p_1 \cdots p_r \, ,$$

where $p_j - 1$ is a power of 2 for each j. The question is then: for which primes p is $p - 1$ a power of 2? It is not hard to show that if $2^l + 1$ is prime, then $l = 2^m$ for some m. Primes of the form

$$2^{2^m} + 1 \, ,$$

$m > 0$, are called *Fermat primes*. Here is a table of the first five:

m	$2^{2^m} + 1$
0	3
1	5
2	17
3	257
4	65537

For $m = 5$, we do not get a prime:

$$2^{2^5} + 1 = 641 \cdot 6700417 .$$

Constructions of the corresponding regular polygons are known for $m = 1, 2, 3, 4$. One for the regular 17-gon is given in [6, Chapter 2], as well as a pretty proof that 641 divides $2^{2^5} + 1$.

Periods

By looking closely at the cyclotomic extension $E_{17}/\mathbb{Q}$, one can give an explicit sequence of quadratic extensions that start with $\mathbb{Q}$ and end at $\mathbb{Q}(\cos(2\pi/17))$. This is done by looking at the *periods* of a primitive 17th root of 1 in E_{17}.

In general let p be a prime, $p > 2$, and let ω be a primitive pth root of 1. We know that

$$\Phi_p(x) = x^{p-1} + \cdots + x + 1 ,$$

and that

$$G := \mathrm{Gal}\left(\Phi_p\right) \cong (\mathbb{Z}/p\mathbb{Z})^\times ,$$

which is cyclic. Let H be a subgroup of G of order h. Define the period

$$\omega_H := \sum_{\beta \in H} \beta(\omega) \in E_p .$$

If $\alpha \in G$ is a generator with

$$\alpha(\omega) = \omega^j ,$$

then α^k is a generator of H, where $hk = p - 1$, and

$$\omega_H = \sum_{l=1}^{h} \omega^{j^{kl}} .$$

For convenience we shall set

$$\omega_k := \omega_{\langle \alpha^k \rangle} = \omega_H .$$

THEOREM 21.3

$$\mathrm{Fix}(H) = \mathbb{Q}(\omega_H)$$

PROOF   First, if $\gamma \in H$, then

$$\gamma(\omega_H) = \sum_{\beta \in H} \gamma\beta(\omega) = \sum_{\beta \in H} \beta(\omega) = \omega_H .$$

Therefore

$$\mathbb{Q}\,(\omega_H) \subset \mathrm{Fix}(H)\,.$$

On the other hand, suppose $\gamma \notin H$. The set $\{\beta(\omega) \mid \beta \in G\}$ is just the set of all primitive pth roots of 1, which is linearly independent over $\mathbb{Q}$. Since

$$\gamma H \cap H = \emptyset\,,$$

it follows that

$$\gamma(\omega_H) = \sum_{\beta \in H} \gamma\beta(\omega) \neq \sum_{\beta \in H} \beta(\omega) = \omega_H\,.$$

Therefore

$$\mathrm{Gal}\left(E_p/\mathbb{Q}(\omega_H)\right) \subset H\,,$$

which implies that

$$\mathbb{Q}\,(\omega_H) \supset \mathrm{Fix}(H)\,.$$

So we have

$$\mathbb{Q}\,(\omega_H) = \mathrm{Fix}(H)\,. \quad \blacksquare$$

Now let $f \in \mathbb{Q}[x]$ be the minimal polynomial of ω_H. It has degree $[G : H]$. For $\gamma \in G$, $\gamma(\omega_H)$ is a root of f. Therefore, as γ runs through a set of representatives of the cosets of H in G, we get a complete set of roots of f. They are of the form

$$\gamma\,(\omega_H) = \sum_{\beta \in H} \gamma\beta(\omega)\,,$$

and are called the h - fold periods of the cyclotomic field.

Examples 21.2

(i) Take $p = 7$. So $\mathrm{Gal}(\Phi_7)$ is cyclic of order 6. It has two nontrivial subgroups, one of order 2 and one of order 3. Let ω be a primitive 7th root of unity. A generator of $\mathrm{Gal}(\Phi_7)$ is

$$\alpha : \omega \mapsto \omega^3\,.$$

The subgroup H of order 2 is generated by α^3. We have

$$\omega_3 := \omega + \alpha^3(\omega) = \omega + \omega^6 = 2\cos(2\pi/7)\,.$$

It satisfies a cubic equation over $\mathbb{Q}$:

$$(\omega + \omega^6)^2 = 2 + \omega^2 + \omega^5$$
$$(\omega + \omega^6)^3 = 3(\omega + \omega^6) + \omega^3 + \omega^4\,,$$

and therefore

$$(\omega + \omega^6)^3 + (\omega + \omega^6)^2 - 2(\omega + \omega^6) - 1 = 0\,.$$

The other roots of $x^3 + x^2 - 2x - 1$ are

$$\alpha(\omega) + \alpha^4(\omega) = \omega^3 + \omega^4 \qquad \text{and} \qquad \alpha^2(\omega) + \alpha^5(\omega) = \omega^2 + \omega^5 \,,$$

corresponding to the cosets αH and $\alpha^2 H$. These are the three 2-fold periods of E_7.

The subgroup of order 3 is generated by α^2, and the 3-fold periods are

$$\omega + \alpha^2(\omega) + \alpha^4(\omega) = \omega + \omega^2 + \omega^4$$
$$\alpha(\omega) + \alpha^3(\omega) + \alpha^5(\omega) = \omega^3 + \omega^6 + \omega^5 \,.$$

They are the roots of the quadratic equation:

$$x^2 + x + 2 = 0 \,.$$

(ii) Take $p = 17$. Then $G := \mathrm{Gal}(\Phi_{17})$ is cyclic of order 16. It has nontrivial subgroups of order 2, 4, and 8. If ω is a primitive 17th root of 1, then

$$\alpha : \omega \mapsto \omega^3$$

is a generator of G. Here is the chain of subgroups with the corresponding tower of quadratic field extensions:

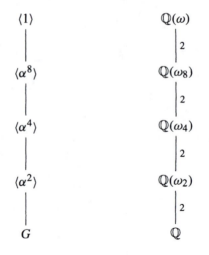

Now

$$\omega_2 = \omega + \omega^9 + \omega^{13} + \omega^{15} + \omega^{16} + \omega^8 + \omega^4 + \omega^2 \,,$$

which is a root of

$$x^2 + x - 4 \in \mathbb{Q}[x] \,.$$

The other 8-fold period is

$$\alpha(\omega_2) = \omega^3 + \omega^{10} + \omega^5 + \omega^{11} + \omega^{14} + \omega^7 + \omega^{12} + \omega^6 \,.$$

Next,
$$\omega_4 = \omega + \omega^{13} + \omega^{16} + \omega^4 ,$$

which is a root of
$$x^2 - \omega_2 x - 1 \in \mathbb{Q}(\omega_2) ,$$

together with
$$\alpha^2(\omega_4) = \omega^9 + \omega^{15} + \omega^8 + \omega^2 .$$

The field $\mathbb{Q}(\omega_4)$ also contains $\alpha(\omega_4)$ and $\alpha^3(\omega_4)$. (These four are the roots of the minimal polynomial of ω_4 over $\mathbb{Q}$). Thirdly,

$$\omega_8 = \omega + \omega^{16} = 2\cos(2\pi/17)$$

is a root of
$$x^2 - \omega_4 x + \alpha(\omega_4) \in \mathbb{Q}(\omega_4) .$$

Lastly, ω is a root of
$$x^2 - \omega_8 x + 1 .$$

By Theorem 21.2, $\cos(2\pi/17)$ can be constructed with ruler and compass, and therefore so can the regular 17-gon. □

Exercises

21.1 Beginning with a unit length, use ruler and compass to construct lengths

   (a) $\sqrt{1 + \sqrt{2}}$;

   (b) $\sqrt[4]{7}$.

21.2 Prove that $4x^3 - 3x - 1/2 \in \mathbb{Q}[x]$ is irreducible.

21.3 Show that it is not possible to construct a regular 9-gon or 11-gon with ruler and compass.

21.4 Prove that if $2^l + 1$ is prime, then $l = 2^m$ for some m.

21.5 Show how to construct a regular hexagon with ruler and compass. Suppose you can construct a regular n-gon, for some n. Give a construction for a regular $2n$-gon.

21.6 Give a ruler and compass construction for a regular 15-gon. Can you generalize this?

21.7 Calculate the periods of the subgroups of $\mathrm{Gal}(\Phi_{13})$. Compute their minimal polynomials over $\mathbb{Q}$.

21.8 What is the minimal polynomial of ω_4 over $\mathbb{Q}$ in Example 21.2(ii)?

Appendix: Mathematica Commands

Here is a list of the *Mathematica* commands introduced in the packages "Groups.m" and "Quartics.m" with their usage statements.

M	`M[a1, ... ,an]` is the permutation which maps 1 to `a1`,..., n to an.	page 32
Dot	`Dot[a,b]`, or a .b, where a and b are matrices or permutations (of the same type), returns the permutation mapping i to a(b(i)). `Dot[X,Y]`, where X and Y are lists of permutations, returns all products a .b as a ranges over X and b over Y. `Dot[a,G]`, or a .G, where a is a permutation and G is a group, returns the list of elements of the left coset. Similarly, G . a returns the right coset.	page 33
Inverse	`Inverse[a]` is the inverse of the permutation or matrix a.	page 33
P	`P[{a1, ... ,am}, {b1, ... ,bn}, ... ]` is a permutation in cycle form.	page 33
Group	`Group[a,b, ... ]` returns the group generated by the elements `{a, b, ... }`.	page 34

Elements	Elements [G], where G is a group, returns the list of elements in G.	page 35
Generators	Generators [G], where G is a group, returns the list of generators defining G.	page 35
Order	Order [a] returns the order of the matrix or permutation a, i.e., the smallest positive integer such that a^n is the identity.	page 35
ChoosePrime	ChoosePrime [p] sets p to be the current ambient prime.	page 45
L	L[{a11, ... ,a1n}, ... {an1, ... ,ann}] is a matrix over $\mathbb{F}_p$, where p is the current prime chosen (see ChoosePrime).	page 46
Orbit	Orbit [a, x], where a is a permutation and x is an integer, or a is a matrix and x is a vector, returns the orbit of x under a.	page 98
Stabilizer	Stabilizer [G, x] returns the subgroup of G fixing x.	page 99
CycleTypes	CycleTypes [G] returns a table of frequencies of cycle types in G.	page 99
ConjugacyClass	ConjugacyClass [G, a] returns the conjugates of a by elements of G.	page 99
Centre	Centre [G] returns the centre of the group G.	page 100
FLTPermutation	FLTPermutation [a], where a is a matrix over $\mathbb{F}_p$, returns the corresponding linear fractional transformation of $P(\mathbb{F}_p)$ as a permutation of $(0, \dots, p-1, \infty)$.	page 100

LeftCosets	LeftCosets[G, K], where G is a group with subgroup K, returns a list of the left cosets of K in G with K itself as the first coset.	page 129
LeftCosetReps	LeftCosetReps[G, K], where G is a group with subgroup K, returns a list of representatives of the left cosets K in G. It consists of the first elements in each coset as given by LeftCosets.	page 129
RightCosets	RightCosets[G, K], where G is a group with subgroup K, returns a list of the right cosets of K in G with K itself as the first coset.	page 129
RightCosetReps	RightCosetReps[G, K], where G is a group with subgroup K, returns a list of representatives of the right cosets K in G. It consists of the first elements in each coset as given by RightCosets.	page 129
Conjugate	Conjugate[a,b] returns a.b.Inverse[a].	page 130
NumberOfConjugates	NumberOfConjugates[G,H], where G is a group with subgroup H, returns the number of subgroups of G conjugate to H.	page 147
CubicResolvent	CubicResolvent[f] returns the cubic resolvent of a quartic f and determines its Galois group.	page 275
QuarticPlot	QuarticPlot[f] plots the pencil of conics associated to a quartic f whose roots are real.	page 276

References

[1] Harold M. Stark, *An Introduction to Number Theory,* MIT Press, Cambridge, 1989.

[2] D.H. Fowler, *The Mathematics of Plato's Academy,* Oxford Univ. Press, Oxford, 1987.

[3] Charles C. Sims, *Abstract Algebra,* Wiley, New York, 1984.

[4] Elmer G. Rees, *Notes on Geometry,* Springer-Verlag, New York, 1983.

[5] Hermann Weyl, *Symmetry,* Princeton Univ. Press, Princeton, 1952.

[6] H.S.M. Coxeter, *Introduction to Geometry,* 2nd ed., Wiley, New York, 1989.

[7] W. Casselman, *Geometry and Postscript,* www.sunsite.ubc.ca/Resources/Graphics/text/www/index.html, 1998.

[8] H.S.M. Coxeter, *Regular Polytopes,* Methuen, London, 1948.

[9] B.L. van der Waerden, *A History of Algebra,* Springer-Verlag, New York, 1985.

Index